# Electric Controls for Refrigeration and Air Conditioning
## Second Edition

Billy C. Langley

DI198567

Prentice Hall, Englewood Cliffs, New Jersey 07632

**Library of Congress Cataloging-in-Publication Data**

LANGLEY, BILLY C.,
    Electric controls for refrigeration and air
conditioning.

    Includes index.
    1. Refrigeration and refrigerating machinery—
Automatic control.   2. Air conditioning—Control.
3. Heating—Control.   I. Title.
TP492.7.L32 1988        621.5'6        87-7017
ISBN   0-13-247511-1
ISBN   0-13-247503-0 (pbk.)


Editorial/production supervision
    and interior design: *Theresa A. Soler*
Cover design: *Photo Plus Art*
Manufacturing buyer: *Lorraine Fumoso*
Page layout: *Audrey Kopciak*

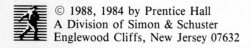 © 1988, 1984 by Prentice Hall
A Division of Simon & Schuster
Englewood Cliffs, New Jersey 07632

All rights reserved. No part of this book may be
reproduced, in any form or by any means,
without permission in writing from the publisher.

Printed in the United States of America


10   9   8   7

ISBN   0-13-247511-1   025
ISBN   0-13-247503-0   025   {PBK.}

PRENTICE-HALL INTERNATIONAL (UK) LIMITED, *London*
PRENTICE-HALL OF AUSTRALIA PTY. LIMITED, *Sydney*
PRENTICE-HALL CANADA INC., *Toronto*
PRENTICE-HALL HISPANOAMERICANA, S.A., *Mexico*
PRENTICE-HALL OF INDIA PRIVATE LIMITED, *New Delhi*
PRENTICE-HALL OF JAPAN, INC., *Tokyo*
SIMON & SCHUSTER ASIA PTE. LTD., *Singapore*
EDITORA PRENTICE-HALL DO BRASIL LTDA., *Rio de Janeiro*

# CONTENTS

## 10   AUTOMATIC GAS-BURNER IGNITION                    136

## 11   WATER HEATING AND COOLING CONTROLS              152

## 12   OIL-BURNER CONTROLS                              167

# PREFACE

The second edition of *Electric Controls for Refrigeration and Air Conditioning* has been expanded to include more types of controls and control systems than the first edition. The workbook and the symbols have also been changed to describe more accurately the modern systems in use.

The purpose of this text is to provide a direct approach to the operation and application of electric controls for both refrigeration and air conditioning systems.

The fundamental concepts employed are applicable regardless of the changes in circuit concepts and design.

This book makes use of a twofold approach conducive to learning:

1. The theory contained in the text assures a thorough understanding of the versatility of each control covered. The material is arranged to provide continuity in learning. Situations are also incorporated to invoke the inquiring minds of students.

2. Practical applications of the fundamentals learned from the text may be directly employed in a student workbook and used in satisfactory laboratory situations. The workbook method has been proven to have student appeal and is a ready reference in later situations.

Even though every effort has been made to relate the principle to the actual application of controls, this is not a conclusive or exhaustive study of the constantly changing control methods employed in the field of refrigeration and air conditioning.

A person who learns the operation and application of the controls and control systems used in refrigeration and air conditioning systems will certainly be in great demand in industry. It is the controls and control systems that cause the equipment to operate as designed.

Upon completion of this study, the student will have the knowledge and confidence necessary to service and install control systems for refrigeration and air conditioning in an efficient manner.

*Billy C. Langley*

# 1 Review of Magnetism, Electrical Circuits, and Transformers

Electricity is the most important form of energy that is used in the world today. If it were not for the availability of electricity, we would not have the convenience of lighting that we enjoy, television and radio would not be possible, and telephones would be nonexistent. Electricity is the form of energy that is used to operate most of our home appliances. Without it, our mode of living would be all but impossible.

The control systems for refrigeration and air conditioning are becoming more complicated with each passing day. A single control can no longer be thought of as being an individual item. Instead, we must consider each control as a component of a more complete system, having a specific reaction to a signal from another component. To understand electric control systems more fully, we must be familiar with the underlying principles upon which they work.

**MAGNETIC FIELDS**   Many of the control devices in use today make use of magnetic fields. The two most popular types of magnets used in the operation of modern control circuitry are permanent magnets and electromagnets.

### Permanent Magnets

Permanent magnets are made from a hardened steel, which when magnetized will hold the magnetic field indefinitely. All types of magnets have north and south poles and all have magnetic fields that surround them.

1

Permanent magnets need no outside power source to make them work as a magnet once they have become magnetized. In control circuits, permanent magnets are used in some thermostats, switches, and valves.

### Electromagnets

In order for electricity to produce magnetism, it must be moving or flowing through a wire or conductor. Therefore, static electricity cannot produce a magnetic field. This is the reason that an open condition anywhere in an electric circuit will prevent the function of all the electromagnetically operated devices in that circuit.

Electrons flowing through a wire cannot pair off with opposite spins, which tend to cancel out any magnetic field produced by their movement. However, when electrons are flowing in the same direction, their magnetic fields tend to add to each other, thus increasing the strength of the magnetic field. The magnetic field around a current-carrying conductor is in concentric circles around that conductor.

When a copper conductor is wound into several turns, the fields around each loop of wire combine to form a magnetic field when current passes through the conductor. One end of the loops is the north pole, and the other end is the south pole. The magnetic strength of the coil is increased when an iron core is inserted inside the coil. These cores are made from a soft iron material. The field-producing force and the permeability (the ability of a material to conduct a magnetic field) of the iron core material both have a definite relationship to the strength of the flux lines. See Figure 1-1. Soft iron has a greater permeability than air.

Electromagnets are used in solenoid valves, relays, gas valves, starters, contactors, thermocouple circuits, and any other device that is automatically opened and/or closed by an electric current. In these types of controls, the electromagnetic field is used to convert the electrical energy to mechanical energy. Therefore, the electromagnetic field is an important principle on which control circuits are based. See Figure 1-2.

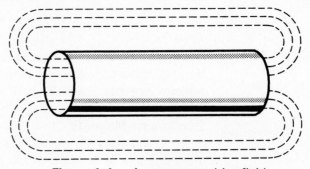

**Figure 1-1.**  A magnet and its field.

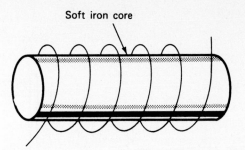

**Figure 1-2.** An electromagnet.

**ELECTRIC CURRENT**

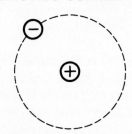

**Figure 1-3.** A hydrogen atom.

Electric current flow is caused by the flow of free electrons in a conductor. Electrons are very small particles within an atom. Everything is made up of atoms. Figure 1-3 shows a single atom of hydrogen. Electric current, which is measured in amperes, is designated by the letter *I*.

The center, or nucleus, of an atom has a positive electrical charge. Electrons, which have a negative electrical charge, revolve around the nucleus. These two oppositely charged particles are strongly attracted to each other. The nucleus of every atom has a total amount of charge equal to the number of electrons revolving around it. Different materials have different atomic structures.

**AMPACITY**

*Ampacity* may be defined as the current-carrying capability of a given conductor. The amount of current that a conductor can carry is determined by the type of material from which it is made, the conductor size, and the thermal properties of the insulation on the conductor. The National Electrical Code (NEC) determines the minimum ampacities (MCA) when rating a given wire size for a given purpose. These MCA ratings are given in tables in the National Electrical Code Manual.

Equipment manufacturers now give ampacity ratings on nameplates on their equipment. Wire size should be selected in conjunction with MCA ratings on the nameplate and NEC guidelines found in the NEC manual.

To prevent nuisance tripping of the circuit breakers and needless blowing of fuses, overcurrent protection allows use of fuses and circuit breakers up to the maximum fuse size (MFS), which is also listed on the equipment nameplate.

**ELECTROMOTIVE FORCE**

The force that causes current to flow is called electromotive force (EMF). EMF is designated by the letter *V* and is measured in volts. EMF can be developed in several different ways. The simplest way is by chemical action on two different kinds of metals. This is known as a cell of battery. Figure 1-4 shows a typical cell, or flashlight battery.

The chemical action of the acid paste causes the materials to give up their free electrons, which then flow from the case through the circuit and back to the carbon rod. The current flows in one direction—negative to positive—and produces direct current electricity.

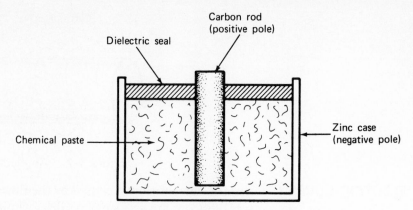

**Figure 1-4.** An electrical cell.

Because energy is removed from this cell and is not replaced, the cell eventually becomes dead. These cells are small and therefore produce a small EMF. The automobile battery, or wet cell, can be recharged from an external source and will last much longer. However, it will eventually use up its source of supply.

The most popular way to produce an EMF is by using the alternating current generator shown in Figure 1-5. In the alternating current generator, a magnetic field is created by an electromagnet that is connected to an external power source. A loop of wire is rotated within the magnetic field. An EMF is developed at the ends as long as the loop of wire is moving and cutting lines of force. To use this EMF, we connect the loop to the outside of the generator by slip rings and brushes. Since the loop rotates within the magnetic field that has a positive force at one end and a negative force at the other end, the EMF will be nega-

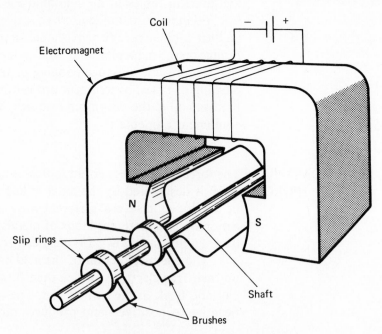

**Figure 1-5.** An ac generator.

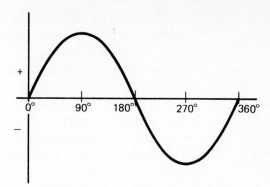

**Figure 1-6.** One ac cycle.

tive during one part of the cycle and positive during the other part of the cycle. The current alternates between positive and negative, and is called alternating current. Figure 1-6 shows one complete alternating cycle.

**RESISTANCE**  Resistance is produced by the circuit through which electric current flows. It may be defined as the opposition to the flow of electric current. Any path through which electric current flows offers some resistance to that flow of current. Resistance is measured in ohms (Ω), named after the German scientist Georg Simon Ohm (1778–1854). Ohm performed experiments with electricity and made some very important discoveries. Resistance is designated by the letter *R*.

A circuit has a resistance of 1 ohm when an EMF of 1 volt causes a current flow of 1 ampere through that circuit.

**OHM'S LAW**  The relationship between current *I*, voltage (EMF) *V*, and resistance *R* is known as Ohm's law and is stated as follows: The current flowing in a circuit is proportional to the voltage and inversely proportional to the resistance. Ohm's law is expressed by the equation:

$$I = \frac{V}{R}$$

which states that current (*I*) equals voltage (*V*) divided by resistance (*R*). Two variations of this equation are

$$E = IR \quad \text{and} \quad R = \frac{E}{I}$$

When any two of these elements are known, the third may be found. In nonmathematical language this formula is interpreted as follows:

As voltage is increased, current increases.
As voltage is decreased, current decreases.
As resistance is increased, current decreases.
As resistance is decreased, current increases.

**SERIES CIRCUITS**  If the same current flows through every part, it makes no difference how many parts or devices there are; the circuit is a series circuit.

When a circuit has a number of resistances connected in series, the total resistance of the circuit is the sum of the individual resistances. An example is shown in Figure 1-7.

Ohm's law applies to the whole circuit as well as to any part of it. In a series circuit the current is the same throughout the whole circuit because it follows a single path. The voltage changes, however, as each resistance in the circuit is encountered. This is known as *voltage drop*. See Figure 1-8.

The current in this circuit is

$$I = E/R = 200/29{,}000 = 0.00690 \text{ amp.}$$

The voltage drop across each resistor is found by:

$$
\begin{aligned}
E_1 &= IR_1 = 0.00690 \times 4000 &&= \phantom{0}27.56 \text{ V}\\
E_2 &= IR_2 = 0.00690 \times 10{,}000 &&= \phantom{0}68.90 \text{ V}\\
E_3 &= IR_3 = 0.00690 \times 15{,}000 &&= \underline{103.35 \text{ V}}\\
& && \text{Total} = 199.81 \text{ V}
\end{aligned}
$$

The sum of the voltage drops is equal to the source voltage. Our answer would have been more accurate if we had calculated the current *I* to more decimal places.

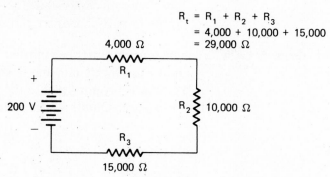

$$
\begin{aligned}
R_t &= R_1 + R_2 + R_3\\
&= 4{,}000 + 10{,}000 + 15{,}000\\
&= 29{,}000 \ \Omega
\end{aligned}
$$

**Figure 1-7.**

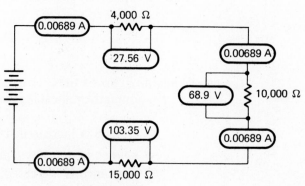

**Figure 1-8.**

Kirchhoff's laws summarize the facts concerning series circuits:

1. The sum of the voltage drops around a series circuit equals the source voltage.
2. The current is the same when measured at any point in the series circuit.

**PARALLEL CIRCUITS**  When more than one component are connected side-by-side to a single voltage source, the components are said to be in parallel. When components are connected in this manner, the total resistance is decreased every time another component is added to the circuit.

A parallel circuit can be compared to a system of pipes carrying water. Two pipes will carry more water than one. Likewise, three pipes will carry more water than two. As more pipes are added, the total resistance to the flow of water is reduced, or decreased. This is illustrated in Figure 1-9. In a circuit with resistances in parallel, the total resistance $R_t$ is found by the equation

$$R_t = \frac{1}{\dfrac{1}{R_1} + \dfrac{1}{R_2} + \dfrac{1}{R_3} + \ldots}$$

$$R_t = \frac{R_1 \times R_2}{R_1 + R_2}$$

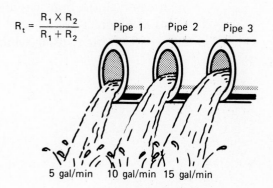

Pipe 1   Pipe 2   Pipe 3

5 gal/min   10 gal/min   15 gal/min

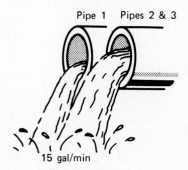

Pipe 1   Pipes 2 & 3

15 gal/min

**Figure 1-9.**

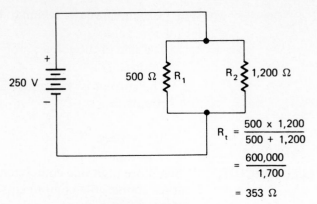

$$R_t = \frac{500 \times 1,200}{500 + 1,200}$$

$$= \frac{600,000}{1,700}$$

$$= 353 \ \Omega$$

**Figure 1-10.**

Figure 1-10 shows an example with just two resistors in parallel.

    With resistances in parallel, the total resistance is always less than the smallest value of any resistor present. This is because the total current is always greater than the current in any individual resistor. Therefore, in the following example the total resistance of the circuit must be one-third of the resistance of a single resistor. If we assign values of 30 ohms to each resistor in Figure 1-11, then

$$R_t = \frac{R \text{ (value of one resistor)}}{N \text{ (number of resistors in network)}} = \frac{30}{3} = 10 \ \Omega$$

    The applied voltage is the same for each branch in a parallel circuit because all the branches are connected across the same voltage source, as in Figure 1-11.

    Therefore, using Ohm's law, the currents are

$$I = \frac{E}{R} = \frac{6 \text{ V}}{30 \ \Omega} = 0.2 \text{ A} \quad \text{across } R_1$$

$$= \frac{6 \text{ V}}{30 \ \Omega} = 0.2 \text{ A} \quad \text{across } R_2$$

$$= \frac{6 \text{ V}}{30 \ \Omega} = 0.2 \text{ A} \quad \text{across } R_3$$

$$I_t = I_{R_1} + I_{R_2} + I_{R_3}$$
$$= 0.2 + 0.2 + 0.2 = 0.6 \text{ amp}$$

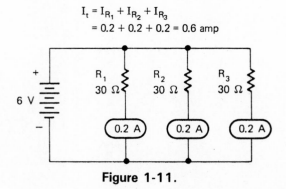

**Figure 1-11.**

The total current flow through the network is the sum of the individual branch currents, or

$$I_t = IR_1 + IR_2 + IR_3 = 0.2 + 0.2 + 0.2 = 0.6 \text{ A}$$

To summarize, we give these laws concerning parallel circuits:

1.  The voltage across all branches of a parallel network is the same.
2.  The total current is equal to the sum of the individual branch currents.

A comparison should be made between these laws and the ones that apply to series circuits.

**SERIES-PARALLEL CIRCUITS**

Series-parallel circuits are a combination of both series and parallel circuits. They can be fairly simple and have only a few components, but they can also have many components and be quite complicated.

In any circuit, there are certain basic factors of interest. From what you have learned about series circuits and parallel circuits, you know that these factors are (1) the total current from the power source and the current in each part of the circuit, (2) the source voltage and the voltage drops across each part of the circuit, and (3) the total resistance and the resistance of each part of the circuit. Once these factors are known, the others can easily be calculated.

When working with either type of circuit, series or parallel, you use only the rules that apply to that type. On the other hand, in a series-parallel circuit, some components are connected in series and some are connected in parallel. Therefore, in some sections of a series-parallel circuit you have to use rules for series circuits, and in other sections you have to use rules for parallel circuits.

Thus, before you can analyze or solve a problem involving a series-parallel circuit, you must be able to recognize which parts of the circuit are series connected and which parts are parallel connected. If the circuit is simple, this is sometimes obvious. However, there are times when the circuit will have to be redrawn in a form that is easier to recognize. See Figure 1-12.

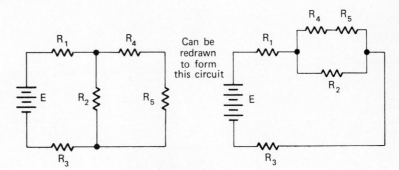

**Figure 1-12.**

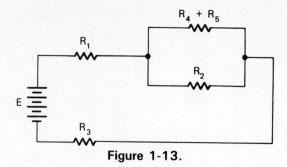

**Figure 1-13.**

There are four steps involved in computing the resistance of a series-parallel circuit:

Step 1 is to redraw the circuit into a form which is easy to recognize.

Step 2 is to combine the series branch circuits. For example, see Figure 1-13.

Step 3 is to find the total resistance of the parallel circuit containing $R_4$, $R_5$, and $R_2$.

$$R_t = \cfrac{1}{\cfrac{1}{R_4} + \cfrac{1}{R_5} + \cfrac{1}{R_2}}$$

The circuit now appears as shown in Figure 1-14.

Step 4 is to compute the series circuit by using Ohm's law.

$$R_t = R_1 + R_2 + R_3 \ldots$$

The circuit now appears as shown in Figure 1-15.

Remember that you usually cannot calculate all the currents of all of the voltages in a series-parallel circuit by using only the total current and applied voltage. You have to work around the circuit load-by-load and branch-by-branch, finding the current through and the voltage across each load or branch before moving on to the next one. As you acquire more experience and practice, you will develop your own shortcuts, which will enable you to eliminate some of the work involved in calculating series-parallel circuits.

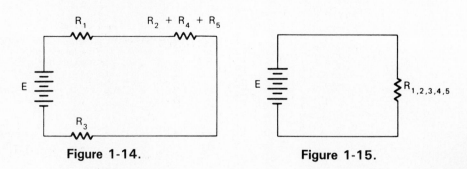

**Figure 1-14.**                    **Figure 1-15.**

**POWER AND ENERGY**  *Power* is defined as the rate of doing work. It is equal to the voltage multiplied by the current. The unit of power is the watt. The formula for figuring power is

$$P = EI$$

By substituting Ohm's law equivalents for $E$ and $I$, we get these additional formulas:

$$P = \frac{E^2}{R} \quad \text{and} \quad P = I^2R$$

Electrical energy is represented by the flow of electric current through a circuit which can be converted to many other forms of energy and is measured in watt-hours. The watt-hour formula is

$$W = PT$$

where

$$W = \text{energy in watt-hours}$$
$$P = \text{power in watts}$$
$$T = \text{time in hours}$$

**INDUCTANCE**  As we learned earlier, a magnetic field contains lines of force. When a current is passed through a straight piece of wire, there is still a small concentration of lines of force, as shown in Figure 1-16. If we form this wire into a coil, we get a stronger concentration of the lines of force. See Figure 1-17.

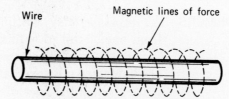

**Figure 1-16.**  Lines of force.

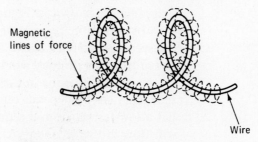

**Figure 1-17.**  Magnetic field in a coil.

The magnetic field created by a current flowing through the coil induces a current in the coil itself as the magnetic lines of force cut across the conductor. This induced current is opposite in polarity to the applied current. When the applied current is increased, the induced current is also increased. This action tends to oppose a change in current. As the applied current starts to build up, it is opposed by the induced current. So, in an inductive circuit, the change in current always lags behind the change in voltage.

**ALTERNATING CURRENT**

To understand alternating current, we must first understand the alternating current (ac) phase. *Phase* is the time interval between the instant one thing occurs and the instant a second and related thing occurs. To find the phase, we divide alternating current into cycles. Each ac cycle takes the same amount of time as all other cycles of the same frequency. Alternating current is divided into 360°, as shown in Figure 1-18.

The ac wave form is a sine wave. As you will note, the current is positive through half the cycle and negative through the other half.

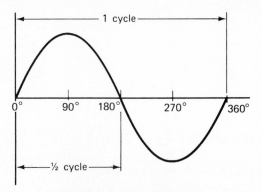

**Figure 1-18.** AC cycle.

**TRANSFORMERS**

A *transformer* may be defined as a device used to transfer electrical energy from one circuit to another.

Basically, a transformer consists of two or more coils wound around one laminated core, so that the coupling between them approaches unity (so that all the lines of flux of one coil will cut across all the turns of a second coil). These devices have no moving parts and require very little maintenance. They are simple, rugged, and efficient. Transformer diagrams are shown in Figure 1-19.

In operation, the primary circuit draws power from the source, and the secondary circuit delivers the power to the load. The power transferred from the primary winding to the secondary winding is determined by the current flowing in the secondary circuit. The current flow in the secondary circuit depends on the power required by the load. If the load has a low resistance and requires a great deal of power, a high current will flow in the secondary circuit. This high current causes a decrease in the EMF of the magnetic field that is necessary for the high current to flow in the secondary circuit. Thus, the transformer regulates the

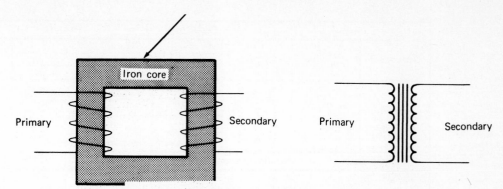

**Figure 1-19.**  Symbolic sketch of a transformer and schematic symbol.

transfer of power from the source to the load in response to the load requirements.

The transformer is used in power transmission to convert power at some value of current and voltage into the same power at some other value of current and voltage. This can be done because with a given voltage in the primary circuit, the secondary voltage depends on the number of turns in the secondary winding as compared to the number of turns in the primary winding.

Careful consideration should be given to the selection of a transformer that is used to power a low-voltage control system. For example, inductive devices such as contactors, relays, solenoid valves, and motors require more power on starting than during steady operation. A transformer delivers the maximum possible inrush current to a load when the transformer impedence equals the load impedence (the total opposition to current flow). The performance of an inexpensive transformer that is properly matched to the load can be equivalent to that of a more costly, unmatched transformer.

## ELECTRICAL SYMBOLS

An electric circuit can be complicated and sometimes difficult to understand. Engineers and technicians have adopted a common set of signs and symbols that are generally used in the drawing of circuits. A drawing using these symbols is called a *schematic.* Equipment manufacturers supply these schematics to aid the technician in locating troubles and in making the necessary repairs. See Figure 1-20.

Schematics show the kinds of parts used and where they are connected in the circuit. The symbols sometimes have a letter identification by which they may be located in the parts list and their type and value found.

As the part is identified on the schematic, you will see a letter close to it, such as *R, C,* or *L.* This letter may also have a subscript. A subscript is a smaller letter located at the lower right side of the letter. It is used to distinguish one part from several others of the same kind.

The parts list is an important part of the description of an electrical circuit. An individual part can be identified by stating its value, its type, and its voltage rating, and by the manufacturer's stock number.

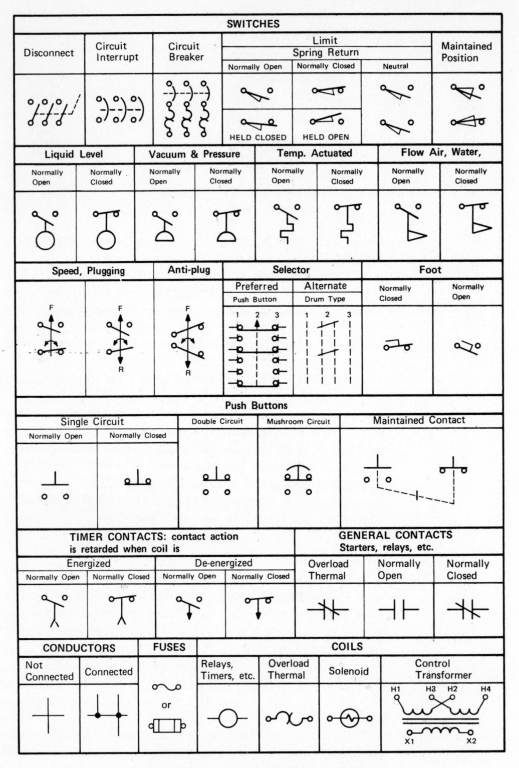

**Figure 1-20.** Graphic electrical symbols.

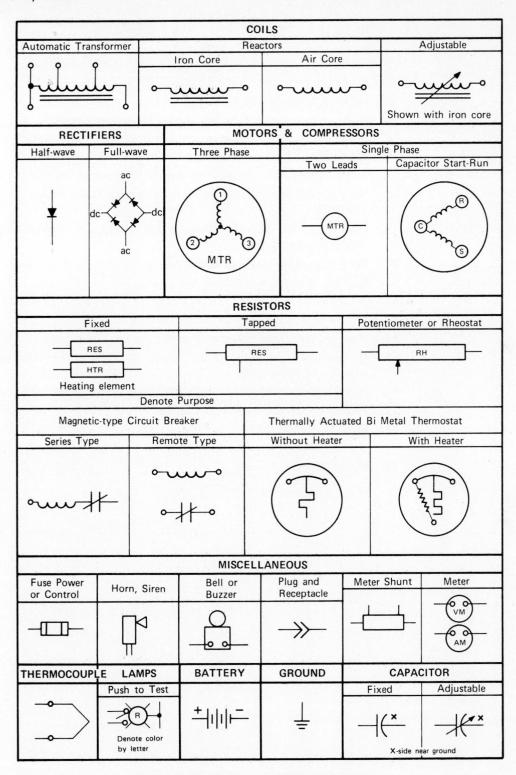

**Figure 1-20** (Cont.).   Electrical diagrams.

## REVIEW QUESTIONS

1. Name two types of magnets used in modern control devices.
2. Name the two poles that all magnets have.
3. Define a permanent magnet.
4. What type of material makes a good permanent magnet?
5. Name the four uses for permanent magnets in modern controls.
6. How can the field strength of an electromagnet be increased?
7. Is the permeability of soft iron greater or less than that of air?
8. What are the two most popular types of magnets used in modern control circuits?
9. Does an increase in current through a coil increase or decrease the magnetic field around that coil?
10. How is an electromagnet made?
11. Give the equation for Ohm's law.
12. Write the formula used to calculate the total resistance in a series circuit.
13. Write the formula for calculating current.
14. State the four laws concerning series and parallel ciruits.
15. In parallel circuits, is the total resistance increased or decreased with the addition of resistances?
16. In a parallel circuit, is the applied voltage the same for each branch?
17. In a series circuit, to what is the sum of the voltage drops equal?
18. Define a transformer.
19. To what coil of a transformer is the line voltage connected?
20. Define power.

# 2

# Transformers

A *transformer* is an inductive device that transfers electrical energy from one circuit to another. When alternating current is applied to the primary coil, a current is induced into the secondary coil. A coupling transformer transfers energy at the same voltage; a step-down transformer transfers energy at a lower voltage and a step-up transformer transfers energy at a higher voltage.

**PURPOSE** The purpose of the transformer used in air conditioning and refrigeration systems is to reduce the line voltage to a low voltage (24 volts) for use in the control circuits. This type of transformer consists basically of two coils of insulated wire wound around an iron core. See Figure 2-1.

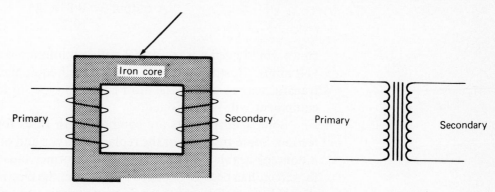

**Figure 2-1.** Symbolic sketch of a transformer and schematic symbol.

The coil connected to the line voltage is called the *primary side* and the coil connected to the control circuit is called the *secondary side*. These coils are step-down transformers; that is, they reduce the line voltage— either 120 volts or 240 volts—to 30 volts.

**RATING**   Transformers are rated in volt-amps (VA) and are of the Class 2 type. This VA rating, listed on the transformer in volt-amps, is an indication of the amount of electrical power that the transformer will provide. The transformer selected must be at least large enough to do the job required of it. This is determined by the amount of load connected to its secondary side. The rating of the transformer may be greater than needed, but if it is smaller than needed, the transformer will only burn out in a short period of time.

**SIZING**   Since most control system components are rated in amperes, it is necessary to know how to compare the VA rating to the ampere draw of the control system. The following formulas are used to make this comparison:

$$\text{Amps} = \frac{\text{VA output}}{\text{secondary voltage}}$$

$$\text{VA output} = \text{amps} \times \text{secondary voltage}$$

Either one of these formulas can be used when selecting a transformer.

The only test that needs to be made is to check the ampere draw in the secondary side of the transformer with all the equipment in operation or with all the equipment in the operating mode that will draw the greatest amount of current. To check the amperage draw, start the unit and check the current draw in the red wire, or the wire that is common to all the load on the secondary side of the transformer.

For example, the amperage draw of the load on the secondary side of the transformer is 0.8 amp. The VA of the transformer would need to be at least

$$\begin{aligned} \text{VA output} &= 0.8 \times 24 \\ &= 19.2 \end{aligned}$$

so we would need a transformer with a minimum output rating of 19.2 volt-amps. Most manufacturers equip their equipment with 40-volt-amp transformers to handle the most common types of loads to which their equipment will be subjected.

Some transformers are equipped with fuses for short-circuit protection. Some fuses are of the replaceable type and others are not. When a nonreplaceable type is used, the tranformer must be replaced when the circuit has been overloaded. Therefore, the electric power to the unit should be turned off when working with wiring connected to the secondary side of the transformer.

There are several reasons for using transformers in control circuits. First, the circuit is much safer than a line voltage circuit would be, thus protecting the customer from electrical shock while protecting the equipment from the greater fire potential of line voltage. Second, the thermostats that can be used with low-voltage control circuits are made from much lighter material than the line voltage type and are, therefore, much more responsive to changes in temperature. They are more economical to produce than line-voltage controls. Third, the low-voltage wiring is greatly simplified when compared to line-voltage wiring.

**PHASING TRANSFORMERS**

Equipment manufacturers sometimes phase transformers for various reasons, but generally phasing is done to increase the power for proper control-circuit energy. *Transformer phasing* assures that all the electrons are flowing in the same direction and is sometimes referred to as *polarizing* transformers.

To phase transformers, wire the primary sides in parallel. See Figure 2-2. Connect one wire from each transformer secondary (B and C) together. See Figure 2-3. Then check the voltage across the other two leads (wires A and D) from the secondary coils. If the voltmeter indicates 48 volts, the transformers are out of phase and the wires must be reversed. See Figure 2-4.

When the transformers are correctly phased, the voltmeter will indicate 0 volts between the two disconnected wires on the secondary side of the transformer. At this point fasten the two wires (A and C) together.

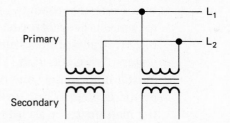

**Figure 2-2.** Transformers wired in parallel.

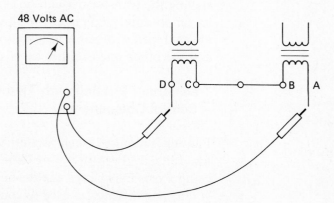

**Figure 2-3.** Checking voltage of improperly phased transformers.

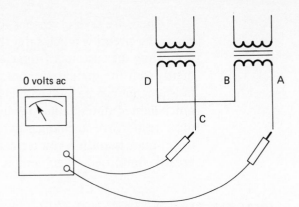

**Figure 2-4.**  Checking voltage of properly phased transformers.

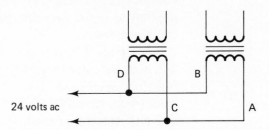

**Figure 2-5.**  Control circuit connected to properly phased transformer.

The transformers are now in phase. The control circuit may now be connected to the two combinations of wires. See Figure 2-5.

From this example, it can be seen that there are two wrong ways to connect the transformer secondaries. One way will result in either one or both transformers being burned out. The other way will result in a voltage that is too high. This high voltage will usually burn out the thermostat immediately and will cause other controls to burn out in a short period of time. Transformers that are supplied with the equipment by the manufacturer are properly phased with marked terminals and are supplied with wiring diagrams. Most transformers are supplied with color-coded electrical leads to aid in proper phasing. The person who is installing the transformer can be certain that proper phasing will result by simply connecting like colors to like colors. Also, transformers that are equipped with screw terminals are color-coded as one screw is brass and the other one is either silver or nickel in color.

### Transformer Phasing with Two Low-Voltage Control Components

Phasing is required when two low-voltage components having their own low-voltage power supply are to be used in the same control circuit. The components may be fan centers or other such devices. There is generally only one thermostat used to control these devices. See Figure 2-6. In this example, the thermostat is presumed to be open. The + and − signs indicate instantaneous voltage polarity. The voltage measured across the

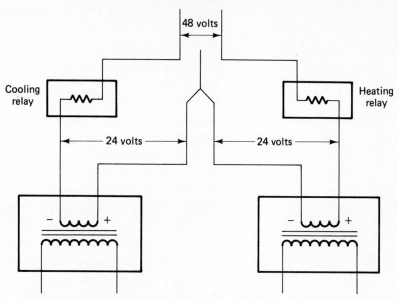

**Figure 2-6.** Phasing control components having separate power supplies.

secondary of either transformer is indicated to be 24 volts, and the voltage across the two wires from the relays is indicated to be 48 volts.

There are two ways to properly phase transformers in this situation. The first method is as follows:

1. Wire the circuit, leaving one of the wires from the thermostat loose at the relay terminal.
2. Check the voltage between the unconnected wire and the relay terminal.
3. If a voltage higher than 24 volts is indicated, reverse the two wires at the control relay.

The second method is as follows:

1. Wire the circuit, leaving the wires disconnected at the thermostat.
2. Check the voltage between one of the wires and the other two in turn.
3. If a voltage greater than 24 volts is indicated, reverse the two thermostat wiring connections at the control relay.

**Transformer Phasing in a Three-Phase Circuit**

When the phasing of transformers in a three-phase circuit is required, the transformer primaries must be connected to the same two legs of the three-phase power. If they are not, exact phasing cannot be accomplished. The secondaries are phased the same way as the single-phase applications described above.

## REVIEW QUESTIONS

1. What type of device is a transformer?
2. What type of transformer is used in refrigeration and air conditioning control systems?
3. How are transformers rated?
4. What can be done to increase the power available for proper control-circuit energy?
5. Write the formula used to calculate the VA output of a transformer.
6. Why are low-voltage control circuits considered safer than line-voltage control circuits?
7. In how many wrong ways can the secondaries be connected when phasing transformers?
8. How can proper transformer phasing be accomplished?
9. What must be done when phasing transformers in a three-phase circuit?
10. How are the transformer secondaries phased in three-phase applications?

# 3

# Magnetic Starters
# and Contactors

In the air conditioning and refrigeration industry, the largest switching load for the control system is without a doubt the compressor motor. The various fan motors, water pumps, and other machinery wired in parallel to the compressor motor also add to the load requirements of the starter or contactor used for controlling these motors and various other loads. There are several methods of controlling the various loads, but we will discuss only the starter and contactor at this time.

**DEFINITIONS** The National Electrical Manufacturers Association defines starters and contactors as follows:

**Starter:** A *starter* is an electrical controller for accelerating a motor from rest to normal speed.

**Contactor:** A *contactor* is a device for repeatedly establishing and interrupting an electric power circuit.

Starters and contactors have many common features. For example, they both have electromagnetically operated contacts that make or break a circuit in which the current exceeds the operating current of the device. Both may be used to make or break voltages that differ from the controlling voltage that is in the control circuit. A single device may be used to switch more than one circuit. Thus, starters and contactors offer the following features as elements of an electric circuit:

1. Isolation between the control circuit voltage and the controlled voltage made complete.
2. High power gain of controlled to controlling power.
3. Ability to control a number of operations from one device with a single control voltage. Also, complete electrical isolation can be obtained between controlled voltages.

**CONTACTOR**   A contactor is used for switching heavy current, high voltage, or both. See Figure 3-1. Contactors generally do not have any overload relays or any of the other features which are commonly found on starters.

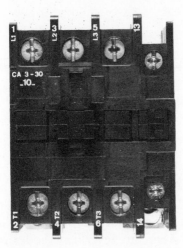

**Figure 3-1.** Three-pole contractor. (Courtesy of Sprecher and Schuh)

**STARTER**   A motor starter may consist of a contactor employed as a means of switching power to a motor. A starter, however, usually has additional components, such as overload relays and holding contacts. These components may also include step resistors, disconnects, reactors, or other hardware required for a more sophisticated starter package for large motors. See Figure 3-2.

*Operation.*   Different manufacturers build starters and contactors with a variety of armatures. Some connect armatures to the relay frame with hinges or pivots, and others use slides to guide the armature. Except for this difference their basic operation is the same; therefore, we will discuss only the hinge type. See Figure 3-3.

The armature is the moving part of the starter or contactor. It is hinged on one end to the frame and has a movable contact assembly on the other end. A spring keeps the armature pulled away from the electromagnet when the coil is de-energized. The metallic armature is easily magnetized by the lines of flux from the electromagnetic coil when it

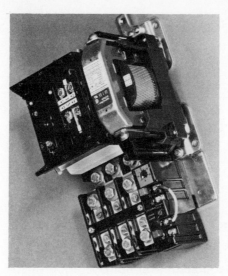

**Figure 3-2.** Three-pole starter. (Courtesy of Sprecher and Schuh)

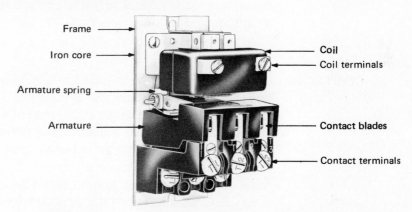

Frame — Iron core — Armature spring — Armature — Coil — Coil terminals — **Contact blades** — Contact terminals

**Figure 3-3.** Contactor components. (Courtesy of Arrow-Hart, Inc.)

is energized. The magnet consists of a coil wound around a laminated iron core. This becomes an electromagnet and pulls the armature toward it when the coil is energized.

**COIL** Coil characteristics depend on the wire and the method of winding. The potential-wound type of coil responds to some value of voltage to pick up the armature. The current-wound type of coil responds to some value of current when energized. Coil terminals connect the coil to the control voltage wiring through a switching device such as a thermostat or pressure control. See Figure 3-4.

The type of service determines the type of insulation used on the coil. Those that are to be exposed to high humidity conditions have a water-resistant type of insulation, whereas those that are not used for this type of service have a different type of insulation. Be sure to use the one recommended by the coil manufacturer.

The voltage rating of the coil must match the voltage of the control circuit. If the control circuit voltage is less than the coil rating, the

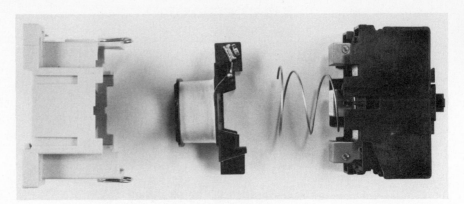

**Figure 3-4**.   Contactor coil. (Courtesy of Sprecher and Schuh)

starter or contactor will not operate properly, if at all. If the control circuit voltage is greater than the coil rating, the coil will burn out in a short period of time, rendering the device useless.

**CONTACTS**  The type of load to be switched is very important when sizing starters and contactors for a particular application. To match the starter or contactor properly to the application, the engineer must consider the different characteristics of inductive and noninductive (resistive) loads.

In an ac circuit with an inductive load, such as an electric motor, the initial inrush of current is always much higher than the normal operating current. Before the motor obtains sufficient speed, the back EMF is very low, allowing maximum current to flow. As the motor gains speed, the back EMF increases and opposes the current flow, reducing it to the normal running current. Starter and contactor contacts, therefore, must not only withstand the normal running current but also the starting inrush of current. A good table of contact ratings lists both full-load and locked-rotor current ratings. See Table 3-1.

Contacts are made of silver cadmium alloy, which gives the highest resistance to sticking. The contacts are bonded to a strong backing member. They operate coolly even at loads of 25% above their ratings.

**TABLE 3–1**
Compressor Motor Current Ratings

| Single-Phase Motor | | Three-Phase Motor | |
|---|---|---|---|
| Compressor Size (Tons) | Motor Current (Amperes) | Compressor Size (Tons) | Motor Current (Amperes) |
| 2 | 18 | 3 | 18 |
| 3 | 25–30 | 4 | 25–30 |
| 4 | 30–40 | 5 | 30–40 |
| 5 | 40–50 | 7½ | 40–50 |

### Pole Configuration

Starters and contactors may have from one to any number of contact poles. Generally the most popular sizes have from one to six poles with one, two, and three poles the most popular. See Figure 3-5. These contacts are used to connect the load to the power source. These devices may be purchased with only the necessary number of poles or they may have extra sets, depending upon the requirements of the system. Any extra poles may be used as auxiliary contacts or they may be left unused.

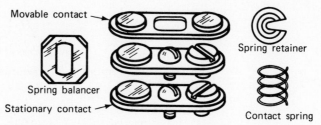

**Figure 3-5.** Set of contacts constituting one pole.

The size of the contacts is determined by the current draw of the load that they are expected to handle. The contacts may be larger than required but they should never be sized smaller than the rated amperage of the load. To do so will cause the contacts to pit and become useless in a short period of time.

Auxiliary contacts are normally used to complete an interlock circuit, such as another contactor connected to a water pump, fan motor, or similar device, or to complete a circuit while the contactor or starter is in the de-energized position. These contacts are generally rated only for pilot duty and are not intended to withstand the heavy current used by the main load.

**OVERLOAD RELAYS** Overload relays are usually mounted on the side of a contactor, thereby completing the definition requirements of a starter. Overload relays may also be conveniently mounted elsewhere in the circuit. They are safety devices used to protect the main load from damage that could possibly be caused by excessive current. In case of an overload condition, the relays may either interrupt the control circuit or they may interrupt one side of a single-phase power line. On three-phase units, the control circuit is always the one to be interrupted.

### Operation

Overload relays operate on the principle that current produces heat. The main load is directed through the relay by use of a resistance wire. If the current exceeds the rating of the resistance wire, additional heat is radiated either to a bimetal switch or to some other type material that is used to hold the contacts together. When sufficient heat is generated, a set of contacts in the control circuit opens and de-energizes the coil

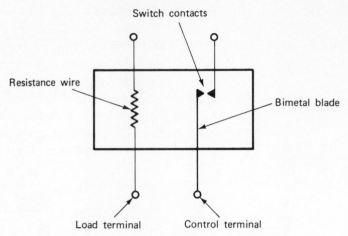

**Figure 3-6.** Representative overload relay.

in the starter or contactor, thereby interrupting the electrical power to the load. The overload may be either manually reset or automatically reset before the starter or contactor will again complete the electrical circuit to the load. See Figure 3-6.

**TWO-SPEED COMPRESSOR CONTACTORS**

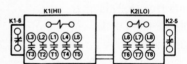

**Figure 3-7.** Two-speed contactor. (Courtesy of Lennox Industries, Inc.)

Many manufacturers are beginning to use two-speed compressors in their units. This requires a different type of starting device, which is a combination contactor having two coils and nine or ten sets of contacts, depending upon the equipment requirements. See Figure 3-7.

Two-speed compressor contactors normally use two auxiliary switches incorporated into the contactors, one on the side of each contactor. These are normally closed (NC) contacts. In most cases, both contactors are mounted on one common base. They have both a mechanical and an electrical interlock to prevent the simultaneous operation of the compressor in both speeds at the same time. The auxiliary contacts provide the electrical interlock. These interlocks must never be prevented from operating. To do so would result in a direct short across the power line, which would cause damage to the contactor and the circuit breaker. Also, any power surges created by this shorted condition could possibly cause damage to the compressor motor windings. The wiring diagrams for each individual unit should be followed so that proper operation of the equipment will be maintained. Control relays (See Chapter 4) are also sometimes used along with these types of contactors to permit the compressor to operate in either the high-speed or the low-speed mode.

**Operation**

To describe the operation of this important control, we use the Lennox HP14 unit with a two-speed compressor.

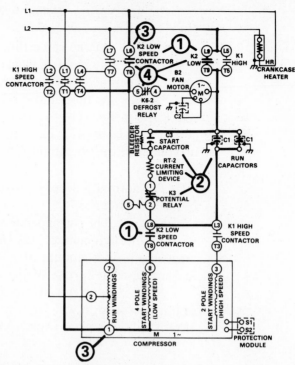

**Figure 3-8.** Compressor starting circuit 1 φ low speed. (Courtesy of Lennox Industries, Inc.)

*Compressor-starting Circuit—single-phase, Low-speed.* See Figure 3-8. The step number corresponds to the circled number in the figure.

**Step 1.** The low-speed compressor contactor (K2) is energized by the timed off control circuit, thus closing contacts L6-T6, L7-T7, L8-T8, and L9-T9.

**Step 2.** To energize the start windings, the start and run capacitors are wired in parallel to provide maximum starting torque for the motor. L2 feeds power through

Low-speed capacitor (K2) contacts L9-T9

Run capacitors (C1)

Start capacitor (C3)

The current-limiting device (RT-2).

The potential relay (NC) contacts (K3).

The low-speed contactor contacts L8-T8 (K2).

The low-speed compressor start windings are energized at the compressor terminal 8.

**Step 3.** The start winding connects the common terminal 1 and completes the circuit to L1 through the low-speed contactor (K2) through contacts L6-T6.

**Step 4.** The fan motor is also energized from L1 through contacts L6-T6 in K2 and the normally closed contacts in the defrost relay (K6-2) and contacts L9-T9 in K2 to L2.

*Compressor-run Circuit—Single-phase, Low-speed.*    See Figure 3-9.

**Step 1.**    At the same time the start windings are energized, the compressor-run windings, which are in series, are powered through the low-speed contactor (K2) contacts L6-T6 and L7-T7.

**Step 2.**    As the compressor comes up to speed, the potential relay coil (K3) is energized by the voltage from the start-motor windings through contacts L8-T8 in (K2). This voltage is usually equal to or above the pick-up voltage of the potential relay. This voltage varies with each compressor model.

The normally closed contacts in K3 open, taking the start capacitor out of the circuit.

**Step 3.**    The run capacitor(s) remain connected to the motor-start winding through contacts L9-T9 and L8-T8 in K2. The run capacitors create the proper amount of voltage shift to improve the power factor and increase the torque of the motor.

*Note:*    If the run capacitor(s) fail, the compressor may not start or if it does start, it will run with a very poor power factor creating high electric bills. The run capacitor(s) provide two functions:

1.  Increasing the starting capacitance when connected in parallel with the start capacitor(s).
2.  Improving the power factor and torque characteristics during the run mode.

*Compressor Starting Circuit—Single-Phase, High-Speed.*    See Figure 3-10. The step numbers correspond to the circled numbers in the figure.

**Step 1.**    L2 feeds power through

High-speed contactor contacts L5-T5 in K1
Run capacitor(s) C1
Start capacitor(s) C3
Current-limiting device RT-2.
Normally closed contacts on the potential relay K3.
High speed contactor contacts L3-T3 in K1.
The high-speed start windings are energized at terminal 3.

**Step 2.**    The start winding connects to the common terminal 1 and completes the circuit through contacts L1-T1 in the high-speed contactor K1.

**Step 3.**    The fan motor is also energized from L1 through contacts L1-T1 in K1, through the normally closed contacts in the defrost relay K6-2, and through contacts L5-T5 in K1.

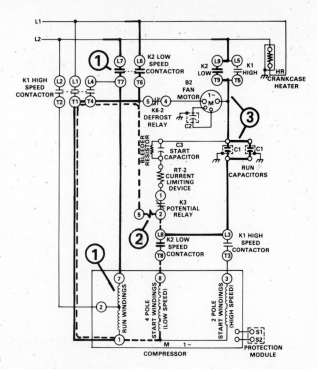

**Figure 3-9.** Compressor run circuit 1 φ low speed. (Courtesy of Lennox Industries, Inc.)

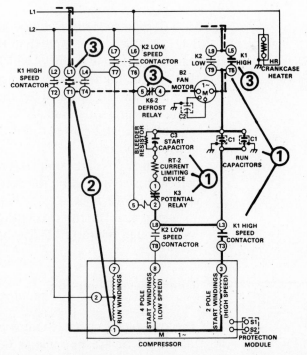

**Figure 3-10.** Compressor starting circuit 1 φ high speed. (Courtesy of Lennox Industries, Inc.)

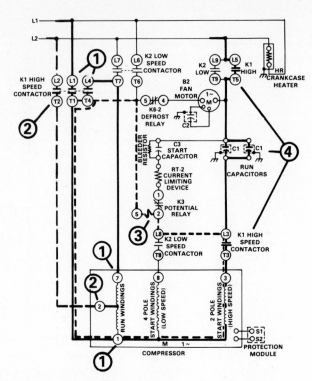

**Figure 3-11.** Compressor run circuit 1 ϕ high speed. (Courtesy of Lennox Industries, Inc.)

*Compressor-run Circuit—Single-phase, High-speed.* See Figure 3-11. At the same time that the start windings are energized, the compressor-run windings, which are wired in parallel, are powered through the high-speed contactor (K1), as follows.

**Step 1.** L1 power is fed through contacts L1-T1 and L4-T4 in K1 to the compressor terminals 1 and 7.

**Step 2.** L2 power is fed through contacts L2-T2 in K1 to compressor terminal 2.

**Step 3.** As the compressor comes up to speed, the potential relay coil (K3) is energized by the voltage from the motor windings through contacts L3-T3 in K1.

**Step 4.** The run capacitor(s) remain connected to the start windings through contacts L5-T5 and L3-T3 in K1.

## Three-Phase Operation

The low-voltage circuits in three-phase units energize the compressor contactors K1 or K2 the same as they do in single-phase units. There are no starting components or run capacitors required or used on three-phase units.

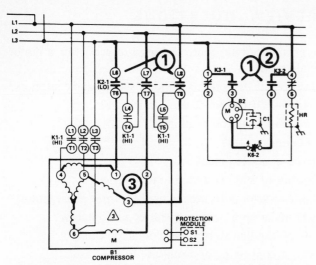

**Figure 3-12.** Compressor circuit three-phase low-speed series Y motor circuit. (Courtesy of Lennox Industries, Inc.)

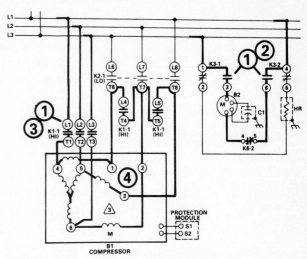

**Figure 3-13.** Compressor circuit three-phase high-speed parallel Y motor circuit. (Courtesy of Lennox Industries, Inc.)

*Compressor Circuit—Three-Phase, Low-Speed Series Y.* See Figure 3-12.

**Step 1.** The K2-1 low-speed compressor contactor and the K3 fan relay are energized by the low-voltage circuit.

**Step 2.** The fan motor B2 is energized from L1 through the normally open contacts in K3-2, the normally closed defrost relay contacts in K6-2, and the normally open contacts in K3-1 to L3. At the same time, the normally closed contacts in K3-1 and K3-2 open, de-energizing the crankcase heater.

**Step 3.** The compressor terminals 1, 2, and 3 are energized through contacts L6-T6, L7-T7, and L8-T8 in K2-1 to form a Y connection to the motor windings for low speed.

*Compressor Circuit—Three-phase, High-speed Parallel Y.* See Figure 3-13.

**Step 1.** The K1-1 high-speed compressor contactor and the K3 fan relay are energized by the low-voltage circuit.

**Step 2.** The fan motor B2 is energized from L1 through the normally open contacts in K3-2, the normally closed contacts in the defrost relay K6-2, and the normally open contacts in K3-1 to L3. At the same time, the normally closed contacts in K3-1 and K3-2 open, de-energizing the crankcase heater.

**Step 3.** The compressor terminals 4, 5, and 6 are energized through contacts L1-T1, L2-T2, and L3-T3 in L1-1.

**Step 4.** The compressor winding terminals 1, 2, and 3 are connected together by contacts L4-T4 and L5-T5 in K1-1 to complete the parallel Y connection to the motor high speed.

## REVIEW QUESTIONS

1.  What is the name of an electrical controller used to accelerate a motor from rest to normal speed?
2.  What force is used to operate the contacts in motor starters and contactors?
3.  How can a contactor be differentiated from a starter?
4.  What causes an overload relay to function?
5.  How are auxiliary contacts used?
6.  Of what are electrical contacts made?
7.  Where are the majority of starters and contactors used?
8.  How many starter or contactor poles are required for three-phase motors?
9.  How many contacts make up a pole of a starter or contactor?
10. Name two methods of winding starter and contactor coils.
11. Name the two ways of connecting an armature to the base.
12. What is the relationship of the control circuit and the controlled circuit when a starter or contactor is used?
13. What is the purpose of a starter or contactor?
14. Depending upon equipment requirements, how many sets of contacts are used in two-speed compressor contactors?
15. What devices are used in contactors to prevent simultaneous operation of a two-speed compressor in both speeds at the same time?

# 4 Magnetic Relays, Thermal Relays, and Compressor Overloads

The relay enjoys the greatest demand in the controls industry. It has many uses in refrigeration and air conditioning control systems. Each year new and imaginative requirements are specified by equipment manufacturers.

**DEFINITION**  A *relay* is a switching device that operates from an electrical input signal to affect the operation of devices in the same circuit or other circuits.

Relays, as studied here, are devices designed especially for the automatic control of one- or two-speed motors in heating, refrigeration, and air conditioning as well as for heating and cooling control. The application of relays will end only with an end to the imagination. It would be almost impossible to cover all applications for relays at one time; therefore, we discuss only a small representation. See Figure 4-1.

**Figure 4-1.** Magnetic relay. (Courtesy of Robertshaw Controls Co., Uniline Division)

**OPERATION**   The operation of magnetic relays is almost the same as the operation of starters and contactors. If necessary, review the operation of starters and contactors. The main difference is the temperature of the relay coil, especially when the relay is enclosed. A coil is limited by the amount of heat it can dissipate in a given time to the ambient air or to the core of the magnet. Increasing the operating voltage causes an increase in temperature.

The outer windings of a coil operate without as much heat as the inner windings because they are closer to the surrounding air and give up more heat. This temperature difference causes hot areas to develop inside the coil, causing deterioration of the coil insulation and shorted windings.

**POLE CONFIGURATION**   Relays may be obtained in almost any type of pole configuration arrangement imaginable. The main arrangements are: normally open (NO), normally closed (NC), single-pole, double-throw (SPDT), double-pole, single-throw (DPST), double-pole, double-throw (DPDT), single-pole, single-throw (SPST), and variations involving these arrangements. For symbols of these arrangements, see Table 4-1.

Some relays have NC contacts and open a circuit when the relay is energized. Circuit diagrams always show relays in the de-energized position.

**TABLE 4-1**
Relay Symbols for Electrical Diagrams.

| Pole Form | Symbol |
| --- | --- |
| SPST, N.O. | |
| SPST, N.C. | |
| SPDT | |
| DPST, N.O. | |
| DPST, N.C. | |
| DPST, N.O. and N.C. | |

**CONTACTS**   Most relay contacts are silver cadmium oxide mounted on beryllium copper blades for longer life and low resistance. Some manufacturers build special models available with gold flash contacts for powerpile systems.

**TYPES OF RELAYS**   There are basically two types of relays used in refrigeration and air conditioning control circuits, electromagnetic and thermal.

### Electromagnetic Relays

When the control circuit is completed and the power is delivered to the relay coil, the electromagnet pulls the armature toward it. As the NO contacts are closed (made), the electrical circuit to the other devices is completed. When the control power to the relay coil is interrupted, the electromagnet loses its power. This permits the armature spring to pull the armature away from the coil, breaking the connection to the output circuit devices.

### Thermal Relays

The thermal relay contacts and terminals are the same as in the electromagnetic relay. The method of pulling the movable contacts against the stationary contacts is the major difference between these two types of relays. The thermal relay uses a bimetal blade with a heater coil wound around it. The heater leads connect to the control-circuit power supply through the control-circuit voltage controller. When the controller demands, the heater warms the bimetallic element, causing it to bend toward the stationary contact to complete the output circuit. When the controller opens the control circuit, the heating element cools and moves away from the stationary contact, breaking the circuit to the external device. See Figure 4-2.

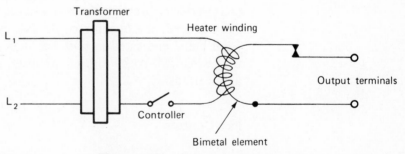

**Figure 4-2.** Thermal relay circuit.

**FAN RELAYS AND FAN CENTERS**

*Fan centers* and *fan relays* are devices that are designed specifically for the automatic control of one- or two-speed fan motors used in heating, refrigeration, and air conditioning systems. They are primarily intended for use with furnace or evaporator fans or blowers. Some original-equipment manufacturers make their own fan centers using a component relay and a transformer.

### Fan Relay

The fan relay may be either an electromagnetic type or the thermal type, depending upon the design of the equipment manufacturer. Its main purpose is to bypass the winter fan control and operate the fan motor for

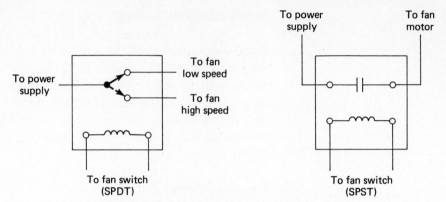

**Figure 4-3.** Typical fan relay schematics.

air conditioning or ventilation. This relay may be purchased alone or with other components inside a housing, known as a fan center, for protection and safety. The contacts of this relay should be heavy enough to withstand the current used by the fan motor. This may be either an SPST or an SPDT relay. The SPST relay has a set of NO contacts; the SPDT relay has a set of NO and a set of NC contacts. They are wired into the circuit according to their use. See Figure 4-3.

   *Operation.*   When the thermostat demands cooling or when the fan switch on the thermostat is turned to the on position, the relay coil or heater is energized, causing the NO contacts to close and the NC contacts to open, directing electricity to the blower motor. When the SPDT relay is used, the opening of the NC contacts prevents the fan motor from being energized in both the high and the low speed at the same time when two-speed motors are used, such as for heating and cooling applications.

## Fan Centers

Fan centers provide the same purpose as fan relays, but they include an integral low-voltage transformer and a terminal board for wiring convenience. They may use either an electromagnetic relay or a thermal relay, and in some cases both are included. The transformer is used to provide low-voltage power for the control system and its components. The thermal relay, when included, delays operation of the compressor to prevent damage due to rapid recycling of the thermostat. The transformer has enough capacity to operate the fan relay, the compressor contactor, and auxiliary equipment controls, providing their combined load does not exceed the transformer output.

   *Operation.*   The operation of the fan center is the same as for the fan relay. The only difference is that the transformer and terminal board are used in fan centers.

**LOCKOUT RELAY**   The lockout relay is used as a compressor motor safety device. The relay coil voltage is the same as the contactor coil voltage, and the contacts may be of the pilot duty type. It is wired into the circuit with its NC contacts in series with the starter or contactor coil. See Figure 4-4.

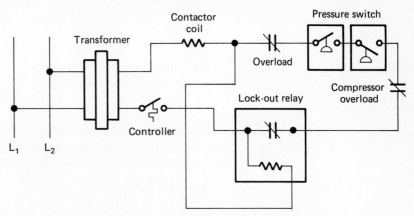

**Figure 4-4.**   Typical lockout relay diagram.

*Operation.*   If any of the other safety devices, such as pressure switches, overloads, or the like, stop the compressor, the lockout relay keeps it from restarting until the unit is reset by turning off the electrical power to it. The electrical power may be turned off either at the main switch or at the controller. When the circuit through the closed contacts is interrupted by one of the safety devices, the voltage is directed through the relay holding coil. This pulls in the armature, opening the contacts to the control circuit. The system is now locked out and will not operate until the system is reset.

**MOTOR-STARTING**   Motor-starting relays are operated either by current or potential (voltage).
**RELAYS**   Current relays can be operated with either an electromagnetic coil or a thermal element, but voltage relays are operated only with electromagnetic coils.

### Voltage (Potential) Relay

This relay may be recognized by its resistance coil wound with very fine wire. The coil of the voltage relay is connected in parallel with the starting winding of the compressor motor.

Potential relays are generally used on high-starting-torque motors, but they may also be used on low-starting-torque motors. The relay contacts are normally closed.

This relay has three terminals, which are used to wire the relay into the circuit. The relay sometimes has as many as six terminals, but the extras are only used as binder or auxiliary terminals, which have noth-

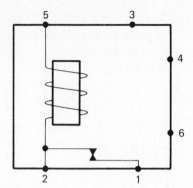

**Figure 4-5.** Symbolic voltage relay.

ing to do with the normal operation of the relay. Terminals 1, 2, and 5 are the operating terminals whereas terminals 3, 4, and 6 are the binding terminals, if these are included on the relay. See Figure 4-5.

When wiring the relay into the system, terminal 1 and the run winding terminal of the compressor motor are connected to the same electric supply line. The starting capacitor is installed in this line. See Figure 4-6. Terminal 5 on the relay and the motor common winding are connected to the same electric power supply line. Terminal 2 on the relay is connected to the motor-starting winding terminal. Terminal 1 is connected to the starting capacitor.

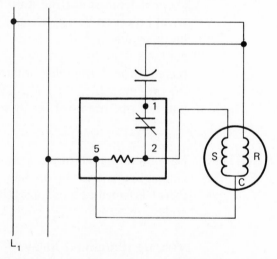

**Figure 4-6.** Typical voltage relay wiring diagram.

Potential relays are easily sized. If the size is not known, start the motor manually and check the voltage between the start and common terminals of the compressor motor while the motor is operating at full speed. See Figure 4-7. Then multiply this voltage reading by 0.75 to find the pickup voltage of the relay.

Table 4-2 shows the specifications of different potential relays. From the table it can be seen that the Mars 67 potential starting relay has a wide range of applications. It can operate with a continuous voltage of 457 volts. The minimum pickup voltage is 295 volts, with a maximum pickup voltage of 315 volts. The maximum dropout voltage is 125 volts.

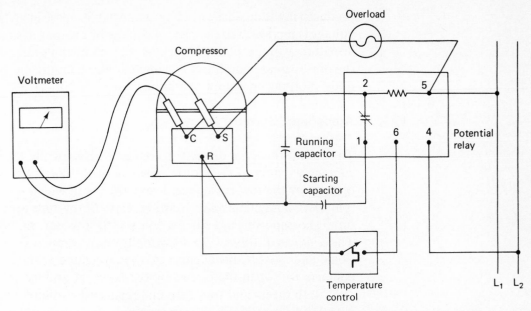

**Figure 4-7.** Checking voltage between start and run terminals.
(Reprinted by permission of Prentice-Hall, Englewood Cliffs, N.J.)

**TABLE 4–2**
Calibration Specifications: Mars—General Electric—Potential Relays

| Mars Relay # | | Contin- uous Volt | | Pick-Up | | Drop Out |
|---|---|---|---|---|---|---|
| | | | | Min. | Max. | Max. |
| Mars 63 | ¼-⅓-½-¾ | 200 | 115v | 139 | 153 | 55 |
| Mars 64 | 1-1½-1¾ | 432 | 230v | 260 | 275 | 120 |
| Mars 65 | ½-¾-1-1½ | 332 | 115v | 168 | 182 | 90 |
| Mars 66 | 3-4-5 | 432 | 230v | 215 | 225 | 120 |
| Mars 67 | 1¾-2-3-4-5 | 457 | 230v | 295 | 315 | 125 |
| Mars 68 | 2-3-4-5 | 502 | 230v | 325 | 345 | 135 |
| Mars 69 | ¾-1-1½ | 378 | 115v | 180 | 195 | 105 |
| Mars 70 | ¾-1 | 253 | 230v | 285 | 305 | 177 |

**Table 4-2.** Calibration specifications: Mars-General Electric—
Potential relays (Reprinted by permission of Prentice-Hall,
Englewood Cliffs, N.J.)

*Operation.*  When power is first applied to the motor, the contacts
of the starting relay are in the closed position. In the closed position there
is an electrical connection between the motor windings. During the initial start-up, there is no back EMF and the maximum current is flowing
through the motor windings. As motor speed increases, a back EMF is

produced in the auxiliary winding causing an increase in voltage. When the voltage is increased to the pickup voltage of the motor-starting relay the armature will "pull in," opening the contact points. At this point the out-of-phase condition is removed, allowing the motor to function under normal operating conditions.

### Amperage (Current) Relay

The coil of this type of relay is connected electrically in series with the running winding of the motor. The coil of the current relay can be recognized by the low-resistance, heavy wire with which it is wound. This is an electromagnetic relay. Starting relays of this type are normally used on $\frac{1}{2}$-horsepower and smaller units with low-starting torque motors. The contacts of this relay are normally open. Amperage relays are positional and must be installed so that the weighted armature carrying the contacts will open the circuit by the force of gravity.

Both three- and four-terminal relays are available, some of which have binding terminals. Three terminal current relays have switch connections from $L$ to $S$ and the coil between $L$ and $M$. See Figure 4-8. Four-terminal relays have switch connections from 3 to $S$ and the coil between 2 and 4 and $M$. See Figure 4-9. Four-terminal relays are often used with three-terminal overloads and provide a convenient method of connecting starting capacitors.

Relays of this type require the use of a motor overload to provide the protection needed in case of high current conditions.

*Operation.*    When power to the motor is first turned on, the contacts of the coil-type current relay are open. The high inrush of current through the electromagnetic coil closes the contacts, thereby producing

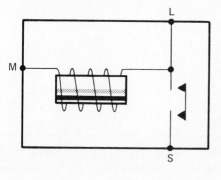

**Figure 4-8.** Symbolic three-terminal current relay.

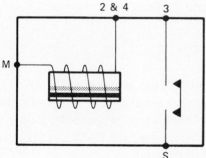

**Figure 4-9.**    Symbolic four-terminal current relay.

the electric phase shift in the start winding required to start the motor. As the back EMF increases, the current decreases, reducing the electromagnetic field and allowing the relay contacts to open and remove the out-of-phase condition. The motor then operates as usual. This occurs when the motor has reached approximately 75% of its full running speed.

Amperage relays must be sized for each motor horsepower and amperage rating. A relay that is rated for too large a motor may not close the relay contacts, which will leave out the much-needed starting circuit. The motor probably will not start under these conditions. A relay that is rated for too small a motor may keep the contacts closed at all times while electric power is applied, leaving the starting circuit engaged continuously. The starting winding will probably be damaged under these conditions. A motor protector must be used with these relays.

### Hot-Wire Starting Relay

The thermal element (hot-wire) type of current relay operates because of heat that is produced by the flow of electric current through the wire. There are two sets of contacts in this type of relay, a set for starting and a set for running. Both sets are normally closed. See Figure 4-10.

*Operation.* When the motor control completes the electric circuit to the motor, the current is automatically supplied to both the running winding and the starting winding at the same time. See Figure 4-11. The current passing through the resistance wire to the main winding causes the wire to heat. This heat causes the bimetal to warp and open the start-

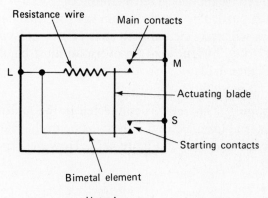

**Figure 4-10.** Symbolic thermal-type current relay.

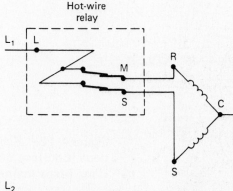

**Figure 4-11.** Wiring diagram of a hot-wire relay. (Reprinted by permission of Prentice-Hall, Englewood Cliffs, N.J.)

ing contacts. The main or running contacts remain closed and the relay is in the normal operating position. Should there be an overload and the motor draw an excessive amount of current, the hot wire gets hotter and causes the bimetal to warp further, opening the running contacts. Both the running and the starting contacts are now open and the compressor motor is stopped. The contacts remain open until the bimetal has cooled down enough to close the contacts. Both the running and the starting contacts close at the same instant and start the motor operating again.

A motor overload is not required when this type of starting relay is used. Some manufacturers do, however, use an internal overload in conjunction with hot-wire relays. These relays are nonpositional, but they must be sized for each motor. If the relay is sized too large, the starting contacts may remain closed for too long a time or not open at all and cause possible damage to the motor windings. If the relay is too small, the relay may stop the motor as if it were in an overloaded condition.

### Solid-State Starting Relays

Solid-state starting relays use a self-regulating conductive ceramic developed by Texas Instruments, which increases in electrical resistance as the compressor motor starts, thus quickly reducing the starting winding current flow to a milliamp level. When the amperage draw is 10 amps or greater, the relay switches in less than 0.35 seconds. This allows this type of relay to be applied to refrigerator compressors without being tailored to each particular system within the specified current limitations. These relays will start virtually all split-phase, 115-volt hermetic compressors up to $\frac{1}{3}$ horsepower. An overload must be used with these relays. See Figure 4-12.

*Operation.* This relay is connected in the motor circuit with the ceramic material installed in the electric line to the starting terminals of the compressor motor. See Figure 4-13. However, the wiring diagram

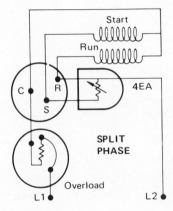

**Figure 4-12.**  Solid-state relay connections (Courtesy of Klixon Controls Division, Texas Instruments, Inc.).

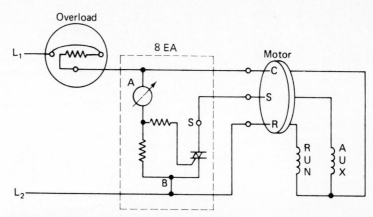

**Figure 4-13.** Solid-state relay connections (Courtesy of Klixon Controls Division, Texas Instruments, Inc.)

provided by the manufacturer should be followed in all instances. The relay is connected in series with the motor-starting winding. When the electricity is applied to the relay, the ceramic material heats up and turns the relay off in approximately 0.35 seconds at 10 amps or greater. This reduces the current flow through the starting winding to a milliamp level until the electric power to the unit is turned off. After the power has been turned off, the ceramic material requires a few minutes to cool down before the next starting cycle should be attempted.

The 8EA relay may not be suitable as a replacement on capacitor-start compressor motors, or for those systems with unusually rapid cycle times.

### Solid-State Hard-Start Kit

The solid-state hard-start kit provides the necessary additional starting torque required to solve starting problems with permanent split capacitor (PSC) motors. Positive temperature coefficient (PTC), or *positive-temperature-coefficient* ceramic materials are used to solve starting problems. The resistances of these materials increase as their temperatures increase. At its anomaly temperature, the resistance increase is very sharp. See Figure 4-14.

*Operation.* The purpose of the PTC start-assist device is to provide a surge of current that lasts only for the period of time needed to start the compressor motor. This additional current is then decreased to allow the motor to run as a normal PSC motor. When the electricity is turned on to the motor, current flows through the start winding and through the parallel combination of the run capacitor and the low-resistance PTC. See Figure 4-15.

The low resistance during the starting phase not only increases the start winding current, but also reduces its angular displacement with the current flowing through the main winding. This is generally an advan-

**Figure 4-14.** Solid-state hard-start kit (Courtesy of Klixon Controls Division, Texas Instruments, Inc.).

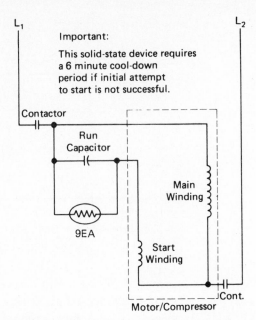

**Figure 4-15.** Solid-state hard-start kit connections. (Courtesy of Klixon Controls Division, Texas Instruments, Inc.)

tage in PSC motors because the phase angle between the starting and running currents is usually greater than 90%.

While the surge of current is providing the necessary starting torque, the current is also flowing through the PTC and heating it to its high-resistance region. The amount of time required for the PTC to heat to its high-resistance state is independent of when the motor starts. Rather, it is a function of the mass of the PTC, the anomaly temperature, its resistance, and the start winding current. When a 230-volt compressor motor using a 9-EA start assist is first energized at a nominal voltage, the 9-EA switching time is 16 electrical cycles, depending on the current flow. Starting the same compressor motor with 25% less voltage increases the switching time of the 9 EA to 32 electrical cycles, providing additional assistance under these more difficult starting conditions.

These switching times are ideal for the PSC compressor motor because the PTC is in its low-resistance state only long enough to overcome the initial inertia of the motor and compressor. When the PTC switching times are longer than the normal motor starting times, the low-resistance PTC effectively shunts out the run capacitor. The excessive on-time retards the motor speed while the motor is attempting to overcome the increasing load.

After the PTC has heated up to its anomaly temperature, its resistance increases to approximately 80,000 ohms and effectively takes itself out of the electrical circuit without the use of an electromechanical relay. As the compressor motor continues to run, the 9 EA draws only 6 milliamps of electrical current. This low-current draw has no effect on the start winding or on the running performance of the motor. As the motor is stopped, the power to the PTC is also turned off and it starts cooling down to the ambient temperature. If the motor should be restarted before the PTC has cooled to its anomaly temperature, the motor will try to start in the standard PSC mode. De-energized time of greater than 1 minute will generally provide sufficient time for the PTC to cool below its anomaly temperature so that the start assist will be available again.

It is recommended that the 9-EA 1 (a solid-state motor starting relay) be used on compressors up to 48,000 British thermal units (BTU) or motors up to 4 horsepower. However, its use is not limited to this size range.

### Positive-Temperature-Coefficient Starting Device

PTC resistor starting devices are used on PSC motors to aid them in starting by applying additional starting torque to the motor. It is not recommended that these devices be used on systems that use a thermostatic expansion valve or on systems that are used in short-cycling applications. They are very simple to install, economical to buy, and have a wide range of application. See Figure 4-16.

The material used for the PTC has a steep-slope, positive-temperature coefficient that has a cold resistance of approximately 50 ohms and a hot resistance of approximately 80,000 ohms.

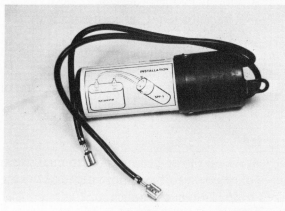

**Figure 4-16.** Positive-temperature-coefficient starting device.

*Operation.* The PTC is wired into the system in parallel with the running capacitor, and it increases the starting torque about 200% to 300%. When the PTC material heats up, it takes the start-assist out of the starting circuit in approximately $\frac{1}{5}$ second. The compressor motor then operates as a standard PSC motor.

When checking this device, allow it to cool down to room temperature; then check its resistance with an ohmmeter. If this reading is very much different from the cold resistance rating, replace the device.

### Hard-Start Kit

A hard-start kit consists of a starting relay, a starting capacitor, and the necessary wiring to install the kit on the unit. The starting relay and capacitor must be of the proper size to prevent damage to the motor. These kits are designed to be used on PSC motors when electrical con-

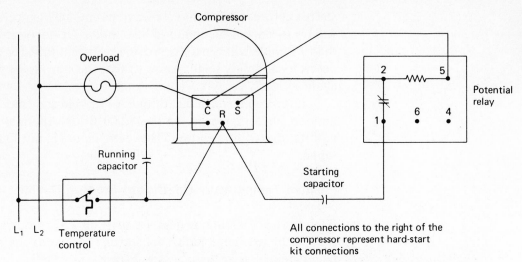

**Figure 4-17** Diagram showing hard-start kit connections (Reprinted by permission of Prentice-Hall, Englewood Cliffs, N.J.)

ditions prevent the compressor motor from starting during normal cycling conditions. They may also be used when rapid cycle conditions are required of the PSC motor. Properly designed, these kits convert the PSC motor to a CSSR (capacitor-start, capacitor-run) motor.

*Operation.*    The kit is installed with the start capacitor between the run and start terminals of the compressor motor, through the relay contacts. As the motor comes up to speed, the contacts are opened, taking the start capacitor out of the circuit. See Figure 4-17.

**HEATING RELAY**    Heating relays are used in the heating unit to energize the heat strips or the gas valve in response to the requirements of the thermostat. The coil is energized by the 24-volt control circuit. The contacts may energize more than one gas valve or step relays for electric heating elements. They may be either bimetal or electromagnetic relays. See Figure 4-18.

*Operation.*    When the thermostat calls for heat, the control circuit energizes the heating relay coil. At this point either the contacts are closed electromagnetically or a resistance heater inside the relay is energized. The normally open contacts are closed, making the circuit to

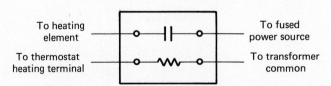

**Figure 4-18.**    Heating relay schematic.

the heating equipment. The heating relay remains in this position until the thermostat is satisfied. When the control circuit is interrupted, the relay will open its contacts, stopping the electric power to the heating equipment. The contact rating of these relays must be heavy enough to withstand the current and voltage of the heating equipment.

**BLOWER CONTROL**    The blower control is used to control the operation of the blower motor on electric heating units. This control will start the blower motor at any time the first heating element in the furnace is energized. These relays have normally open contacts. When they close, the blower motor is energized, moving air to the conditioned space. These line-voltage contacts are opened and closed by the low-voltage control circuit. See Figure 4-19.

*Operation.*    The low-voltage part of the blower control has an electronic sensing circuit, which is connected to a current-sensing loop mounted on the outside of the control. The two terminals of the electronic circuit are connected across the secondary of the transformer. See Figure 4-20. The wire from the hot leg connected to the heating element passes through the sensing loop. A current of 15 amps or more passing

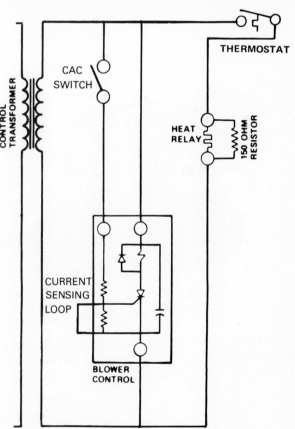

**Figure 4-19.**   Blower control. (Courtesy of Lennox Industries, Inc.)

**Figure 4-20.**   Blower control wiring diagram. (Courtesy of Lennox Industries, Inc.)

through the sensing loop activates the electronic circuit, which then closes the line voltage contacts of the blower control, energizing the blower motor. The heating element draws 20 amps of current any time it is energized, insuring that the blower will be running when the element is on.

**TIME-DELAY SEQUENCING RELAYS**

Time-delay sequencing relays are bimetal-type relays that are normally used with electric heating units. They are used in furnaces equipped with several heat strips for heating the building. They are used to reduce the input of current to the unit during start-up by staging, or sequencing, the elements to come on at definite time intervals. This procedure may be accomplished by using 24-volt heaters in the relays or a combination of 24 volts and line voltage to the heaters, depending upon the manufacturer's design and requirements. Relays may be selected that have different time-delay characteristics that provide the desired time delay between the energizing of each element on demand from the thermostat, or they may be selected so that the first one energized has a low-voltage heater and the remaining relays have line-voltage heaters, which are energized from a second set of contacts in the relay. They may be either single relays or stacked relays, depending upon the application. See Figure 4-21.

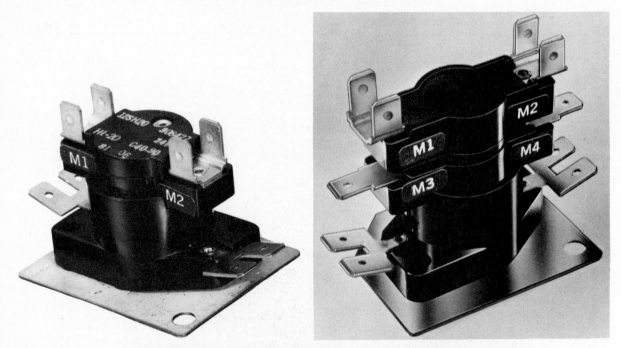

**Figure 4-21.**   (a) Single and (b) stacked time-delay relays.
(Courtesy of White-Rodgers Division, Emerson Electric)

*Operation.*   When the first relay has a low-voltage heater and the remaining relays have line-voltage heaters, the operation is as follows: When the thermostat demands heat, the first relay heater is energized.

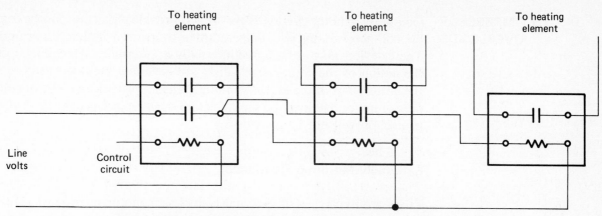

**Figure 4-22.** Series-wired time-delay sequencing relay wiring diagram.

After the specified amount of time has passed, the relay closes its contacts, energizing the first heating element and directing line voltage to the heater of the second relay. After the prescribed amount of time has passed, the second relay closes its contacts, energizing the second heating element and directing line voltage to the third relay heater. This sequence continues until all the heating elements are energized. See Figure 4-22.

When the thermostat is satisfied, the low voltage to the first relay is interrupted, allowing the heater to cool to the point that the contacts open and stop the electricity to the first heating element and to the heater of the second relay. After sufficient time has passed for the second relay heater to cool, the contacts open, stopping the electricity to the third relay, and so on until all the relays have been de-energized.

When all the relays are equipped with low-voltage heaters, the operating sequence may be as just described or as follows: When the thermostat demands heat, all the heaters in all the relays are energized at the same time. However, because each has a different time interval before the contacts are closed, the heating elements will be sequenced automatically. See Figure 4-23.

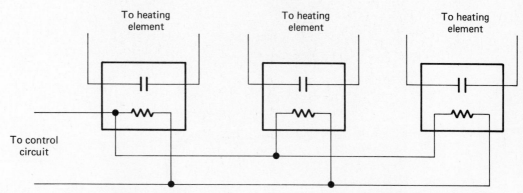

**Figure 4-23.** Parallel-wired time-delay sequencing relay wiring diagram.

**COMPRESSOR OVERLOADS**

Compressor motor overloads are used to provide protection for the compressor motor during overtemperature or overcurrent conditions or both. These devices may be mounted internally or externally, depending upon the design of the compressor. They are generally mounted near or in the hottest part of the motor winding. When replacing any type of compressor motor overload, be sure to use an exact replacement or the proper protection will not be provided.

### Externally Mounted Overloads

There are two types of external compressor motor overloads: (1) the bimetal operated type and (2) the hydraulic operated type. See Figure 4-24. The bimetal types operate in response to both the temperature of the motor winding and current to the motor. The hydraulic type operates in response to the temperature of a hydraulic fluid being heated by the current flowing through the overload. These overloads are manufactured with two, three, or four electrical terminals.

*Bimetal-Type Overload.* The bimetal types are designed with the following two different methods of stopping the compressor motor in case of an overload: (1) the bimetal breaks the line voltage to the motor

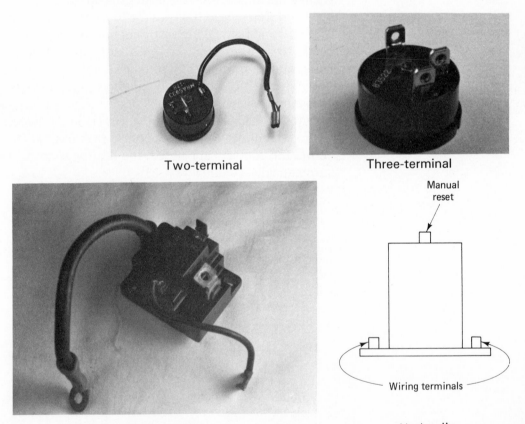

Two-terminal                    Three-terminal

Four-terminal                    Hydraulic

**Figure 4-24.** External compressor overloads.

when an overloaded condition is sensed, and (2) the bimetal breaks the control voltage. This causes the contactor or starter to open, which then interrupts the line voltage to the compressor motor and stops the compressor motor.

*Operation.*   When the compressor motor is first energized, there is a slight time delay before the overload opens the electrical circuit. This is because the bimetal is cool and it takes a certain amount of time before it has warmed to the opening temperature. By the time it has warmed to this temperature, the compressor amperage will have, under normal conditions, dropped to the normal operating range. The overload will remain in this position either until the motor overheats or an excessive amount of current is drawn by the motor or until the compressor is cycled off under normal conditions. If the overload senses a sufficient increase in heat caused by overcurrent or some other type of overload, it will open its contacts, interrupting the electrical circuit to the compressor motor. The compressor will then stop until the overload has cooled sufficiently to allow the motor to restart or until the overload is manually reset, if it is of the manual reset type. This cycle will continue until the overload condition has been corrected or until either a fuse is blown or a circuit breaker is opened. The cause for the overload must be found and corrected or the condition will only get worse.

*Hydraulic Type Overload.*   This type of overload is generally mounted away from the compressor in a control panel. It may be of the manual or the automatic reset type, depending upon the equipment design requirements. It operates on current draw to the compressor motor only and is used only on larger compressor motors. It uses the temperature of a hydraulic fluid to open and close the contacts.

*Operation.*   When the compressor motor starts, the hydraulic fluid is cool and it takes a few seconds before the fluid will sense an overload condition. Therefore, a slight time delay is provided to prevent nuisance shutdowns. All the current to the motor is directed through a heater, which warms the hydraulic fluid. This warming of the fluid causes an increase in the fluid pressure which in turn causes a set of contacts to open interrupting the electrical circuit to the compressor. After the hydraulic fluid has cooled down, the overload may be manually reset or reset automatically, as required, before the compressor motor can be restarted. This type operation will continue until the cause of the overload has been corrected.

**Internal Overloads**

Internal overloads may be of the line-break type or the thermostatic type. Both types of internal overloads are located precisely in the heat sink portion of the motor windings and protect the motor from excessive temperatures and excessive current draw. See Figure 4-25.

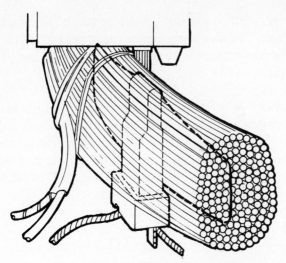

**Figure 4-25.** Internal overload location. (Courtesy of Lennox Industries, Inc.)

*Operation.*   The line-break type interrupts the electric power line to the common connection of the motor winding if either of these conditions occurs. See Figure 4-26. The internal overload protector is also mounted in the motor windings. However, the switch interrupts the control circuit to the contactor or starter. The electrical connections are shown in Figure 4-27. If either of these overloads become faulty, the compressor must be replaced.

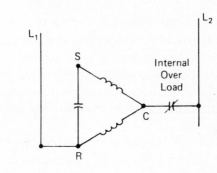

**Figure 4-26.**   Line break internal overload connection. (Reprinted by permission of Prentice-Hall, Englewood Cliffs, N.J.)

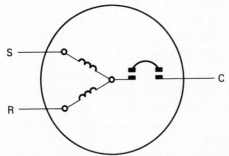

**Figure 4-27.**   Internal overload connections.

**SHORT-CYCLE-TIME DELAYS**    It has been determined that heat is the greatest enemy of electric motor winding insulation. Any heat damage that occurs to the winding is permanent and cannot be reversed. It is, therefore, irreversible and cumulative. Heat in a compressor motor can be caused by the compressor motor operating on low voltage, single phasing, voltage unbalance, or overload conditions that are caused by the compressor starting before the refrigerant pressures have equalized. These types of controls provide the required protection by sensing these conditions and stopping the compressor motor before damage can occur. Time-delay controls react much faster than thermal type controls, which are slower to sense and operate because the heat must first be generated somewhere in the motor and then be conducted to the sensing device.

### Delay-On-Break Timers with Relay Outputs

*Delay-on-break timers with relay outputs* prevent short cycling of the compressor caused by voltage or thermostat interruptions. The "off delay" begins immediately when the interruption occurs. The circuit terminates through four $\frac{1}{4}$-inch quick disconnect terminals. See Figure 4-28.

The internal 1-amp relay operates on signal from the 24-volt control circuit, which drives the control relay direct. These relays are factory preset with a choice of delays of 3, 4, and 5 minutes available.

*Operation.* With either a voltage interruption or an interruption by the thermostat, the relay will drop out. After the specified amount of time has passed, the compressor is re-energized and a start is attempted. If the situation has been corrected, the compressor starts and operates as usual. If the situation has not been corrected, the compressor motor is again stopped by the protector for the predetermined period of time.

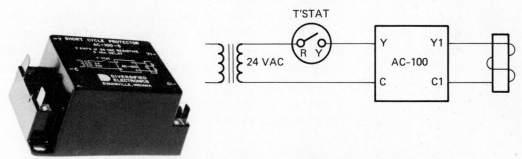

**Figure 4-28.** AC-100 series delay-on-break timer with relay output. (Courtesy of Diversified Electronics, Inc., Air Conditioning/Refrigeration Division)

### Delay-On-Break Timers with Relay Output and Random Restart

*Delay-on-break timers with relay output and random restart* are solid-state timers that prevent short-cycling of compressors caused by thermostat interruptions or electrical power failures. See Figure 4-29. In addition, these timers prevent circuit overloading in multiple-installation systems by providing a random restart when the power is restored. These relays are equipped with a 5-minute time-delay period.

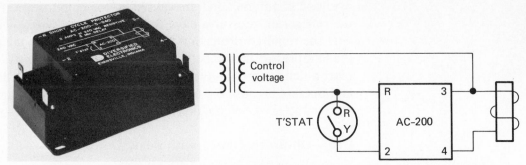

**Figure 4-29.** AC-200 Series delay-on-break with relay output and random restart after power failure. (Courtesy of Diversified Electronics, Inc., Air Conditioning/Refrigeration Division)

*Operation.* When the thermostat opens, the control relay drops out and the 5-minute delay period begins. Any subsequent opening and/or closing of the thermostat contacts during the delay period does not lengthen the timing sequence. When the thermostat contacts are closed after the delay period, the control relay immediately picks up. If the thermostat is opened at the end of the delay period, the control relay remains dropped out until the thermostat contacts close again. Under normal operating conditions, the compressor off-time is generally longer than 5 minutes. Under these conditions a timer will not lengthen the compressor off-time. On applications when the compressor off-time is less than 5 minutes, the addition of a timer increases the compressor off-time to 5 minutes.

When a power outage occurs, the control relay drops out immediately. The 5-minute time delay starts when the power is restored. When the power is restored, the control relay pulls in, provided there are no other interruptions present and the thermostat contacts are closed. Any voltage interruption during the timing period automatically resets the timer, and the compressor starts 5 minutes after the last interruption upon demand from the thermostat.

**ELECTRIC HEAT RELAY**

*Electric heat relays* are designed to control conventional on-off electric heating elements and to control operation of the fan in an electric furnace. They may be used on furnaces that use either line-voltage or pilot-duty limit controls. They are the same as the contactors and relays discussed earlier. These controls must be selected with high enough contact rating to switch the required load properly. Electric heat relays may be wired to control any combination up to four heating elements. See Figure 4-30.

*Operation.* Electric heat relays are energized by a thermostat, which closes two sets of NO contacts. One set is used to energize the fan motor and a heating element or elements. The other set is used to energize the other heating element or elements. In this manner, the fan is always energized as long as the thermostat is demanding heat. All the elements are energized at the same time.

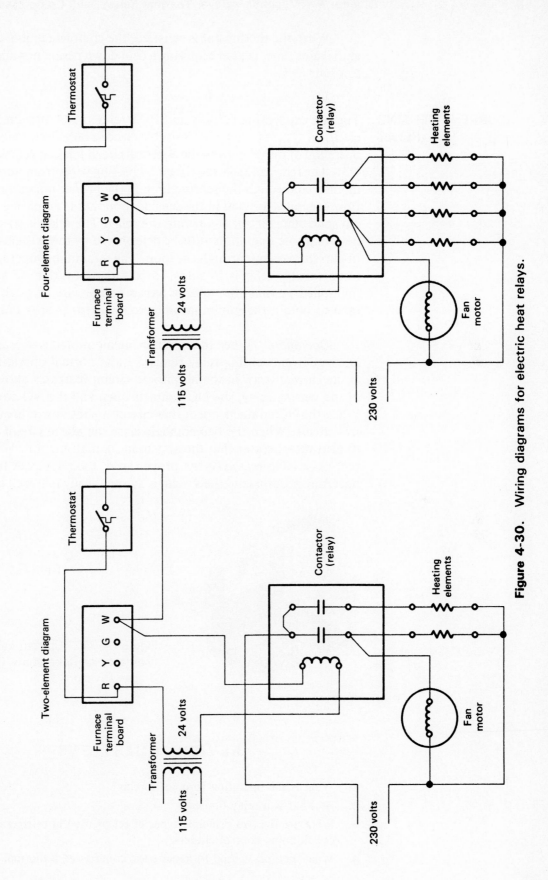

**Figure 4-30.** Wiring diagrams for electric heat relays.

**57**

When the thermostat is satisfied, the complete unit is de-energized at the same time, neither requiring a cool down period nor allowing drafts to occur.

**CURRENT-SENSING RELAY**

The current-sensing relay is an SPDT relay that is designed for use on electrical systems when it is desirable or necessary electrically to separate operation of the relay from the electrical circuit it is controlling. See Figure 4-31. The relay controls the circuit by passing wires from the circuit being monitored through the sensing loop in the proper polarity. The current flow generates a signal in the sensing loop that is amplified by an internal solid-state circuit to operate the relay. The relay is used to operate devices whose operation must be determined by the existence or absence of current passing through the loop. The minimum amperage necessary to operate the relay is 15 amps. Circuit loads less than 15 amps may be multiplied by making multiple passes of the wire through the sensing loop to obtain the minimum ampere turns for proper operation.

*Operation.*    When the circuit being monitored is operating normally, the current-sensing relay remains in its normal operating position. If the current draw in the monitored circuit increases above the rating of the sensing relay, the NC contacts open and the NO contacts close. When the NC contacts open, that circuit is shut down through the control circuit. When the NO contacts close, an alarm circuit is energized to alert the operator that the system has a malfunction. When the current draw is decreased to the proper level, the relay goes to its normal operating position and the system is automatically put back in operation.

**Figure 4-31.** Current-sensing relay. (Courtesy of Robershaw Controls Co., Uniline Division)

## REVIEW QUESTIONS

1. What is the operating force of a relay?
2. By what is a relay limited?
3. What are the two principle types of relays used in refrigeration and air conditioning control circuits?
4. What method is used to actuate the contacts of a thermal relay?

5. What devices are used for automatic control of the operation of one- or two-speed fan motors used in heating, refrigeration, and air conditioning systems?

6. What is the difference between a fan relay and a fan center?

7. What type of device is a lockout relay?

8. In relationship to the compressor motor, how is the coil of the voltage starting relay connected?

9. In relationship to the compressor motor, how is the coil of the current-starting relay connected?

10. What force causes the hot-wire starting relay to operate?

11. How do solid-state starting relays operate?

12. How is a solid-state starting relay electrically connected to the motor?

13. On what type of motors are PTC starting devices used?

14. What device may be installed on PSC motors when electrical conditions prevent the compressor motor from starting during normal cycling conditions?

15. What device may be used on heating systems to energize the heat strips or the gas valve in response to the thermostat requirements?

16. Are the contacts of a blower control normally open or normally closed?

17. What type of relays are normally used on electric heating units to control operation of the heat strips?

18. Name the types of externally mounted compressor motor overloads.

19. Where are internal motor overloads located?

20. What type of relay is used on systems when it is desirable or necessary to separate electrically operation of the relay from the electrical circuit that it is controlling?

# 5 Solenoid Valves, Reversing Valves, and Compressor Unloaders

Solenoid valves, reversing valves, and compressor unloaders are used in refrigeration and air conditioning to direct, channel, or directly affect the flow of fluids.

The automatic control of refrigerants, brine, gas, or water depends frequently on the use of one or more of these valves. They may be used individually or together in a system to compliment each other and to provide an ultimate in control systems.

**DEFINITIONS** The following are the accepted definitions of these various components and controls.

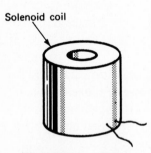

Solenoid coil

**Figure 5-1.** Solenoid coil.

**Solenoid Valves:** A *solenoid valve* is a device made by putting two separate devices into one control. A solenoid is a coil of wire, which—when carrying an electric current—has the characteristics of an electromagnet. See Figure 5-1. When this coil is attached to the stem of a valve, a plunger is pulled into the coil, providing the desired action of the valve. Hence it is called a solenoid valve.

**Reversing Valve:** A *reversing valve* (four-way valve) provides four flow paths—two at a time. Two separate flows can be diverted in two directions by virtue of the valve operation.

**Compressor Unloader:** A *compressor unloader* loads or unloads compressor cylinders by allowing discharge gas to bypass to the suction side of the cylinder through a bypass port, thus controlling the capacity of the compressor.

**SOLENOID VALVES**     Of all the controls in the refrigeration and air conditioning industry, solenoid valves are probably the most used. Because they are electrically operated and perform many functions of a manually operated valve, they are desirable for automatic operations.

These valves can be made to perform many functions merely by turning on or off an electrical circuit, which is usually accomplished by a thermostat. When the electrical circuit to a solenoid coil is completed, an electromagnetic field is created, which pulls the plunger into the center of the coil. See Figure 5-2. By attaching this plunger to a valve stem, the valve can be made to open or close upon energizing or de-energizing the electrical circuit.

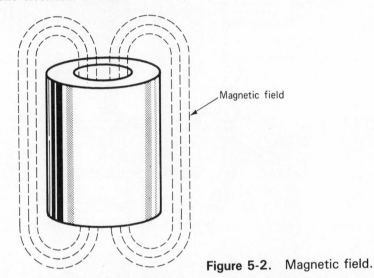

Magnetic field

**Figure 5-2**.   Magnetic field.

There are two general types of solenoid valves: (1) the direct-acting (Figure 5-3) and the pilot-operated (Figure 5-4).

*Operation.*    In the direct-acting type, when the circuit is completed to coil A, the energized coil pulls plunger B upward lifting valve disc C from valve seat D. This allows the fluid to pass through the valve until such time as the circuit is broken to coil A. The plunger B then drops, and valve C rests on seat D, shutting off the flow of gas. Valve spring E assures positive closure.

Valves that are equipped with manual operation have a knob on the bottom of the valve body. In case of power failure the valve may be opened by sliding the instruction sleeve off the knob and pushing the knob upward with the fingers. Then the knob is rotated one-quarter turn in either direction to hold valve C in the open position.

To close the valve manually, rotate the knob one-quarter turn and the manual operator will return to its normal position. The force of valve spring E returns the valve to the fully closed position.

*Note:*   On manual operation, the control does not return to the automatic operation when the power is restored.

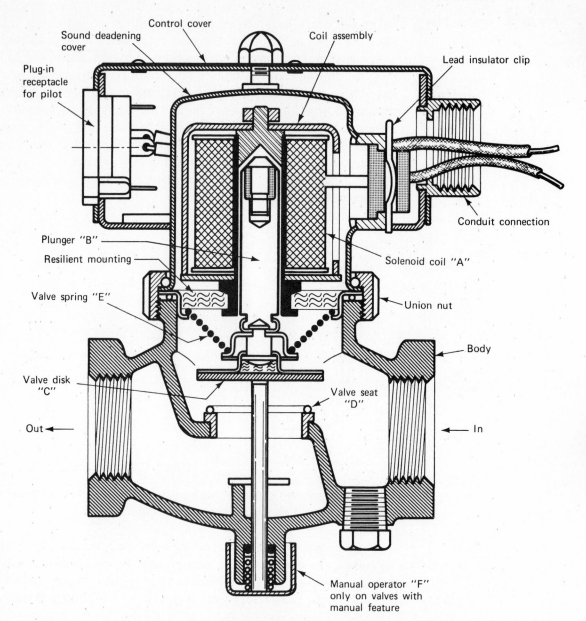

**Figure 5-3.**   Direct-acting solenoid valve.

Pilot-operated solenoids are usually made in the larger sizes. The plunger does not directly open the main valve seat in this type of valve. When coil A is energized, plunger B, which has two seats, moves from seat C to seat D. Pilot port E is thus closed and bleed port F is opened, allowing the fluid to escape to the outlet side of the valve. The pressure is now the same on both sides of diaphragm G, allowing spring H to open the main seat and allow the fluid to flow straight through the valve.

All solenoid valves are made on these basic principles. There are, however, a few exceptions in the mechanical construction of these valves. Some examples are the lever valve, the general-purpose two- or three-way valve, and the four-way valve. The proper valve should be chosen for the specific installation.

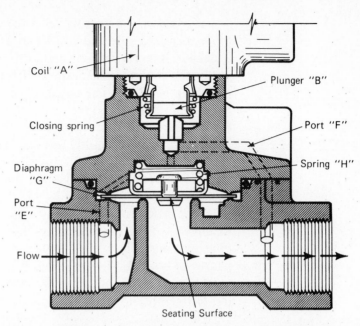

**Figure 5-4.** Pilot-operated solenoid valve.

For the solenoid valve to function properly, the following general rules should be kept in mind when the valve is installed.

1. Follow the arrows on the valve body, which indicate the proper direction of flow through the valve.
2. It is advisable to use new pipe, properly chamfered and reamed, when making connections. Be careful when using pipe dope.
3. Be sure the ambient air temperature immediately surrounding the solenoid valve does not exceed 125°F (51.5°C).

**REVERSING VALVES** Reversing valves are designed for various nominal tonnage capacities and for the automatic operation of heat pump air conditioning systems. Automatic temperature controls are used in addition to the reversing valves. They are also used on commercial refrigeration systems that use heat for defrosting the evaporator coils. They should be sized according to the manufacturer's specification. These valves are hermetically constructed and are pressure-differential operated. See Figure 5-5.

**Figure 5-5.** Reversing (four-way) valve. (Courtesy of Ranco Controls Division)

*Operation.*     Reversing valve operation is controlled by an energized or de-energized solenoid coil secured over a three-way pilot valve with a locknut, integral with the main valve. The coil may be energized either during the cooling cycle or during the heating cycle. One is no more correct than the other. The choice depends upon the desires of the equipment manufacturer. In the discussion that follows, we will assume that the coil is energized during the heating cycle.

These valves instantly reverse running system refrigerant pressures and operate wholly on pressure differential between the high and low sides of the refrigeration system under full pressure within its listed capacities.

The refrigerant gas path is schematically diagrammed through the main valve. This diagram shows the sliding port at a position over two tube openings as it transfers both refrigerant coils between the operating phases of cooling, de-icing, and heating.

The solenoid coil is not energized in the normal cooling cycle and the refrigerant flows through its conventional cycle. See Figure 5-6.

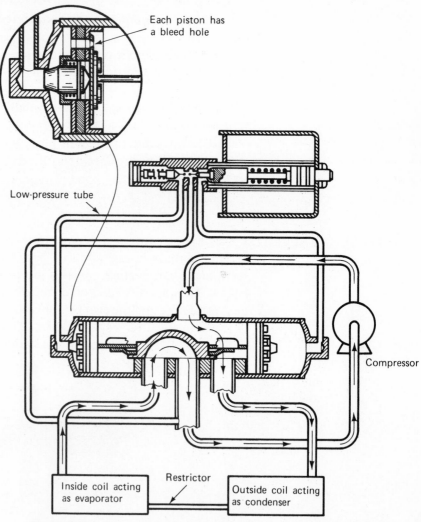

**Figure 5-6.**     De-energized solenoid coil. (Courtesy of Ranco Controls Division)

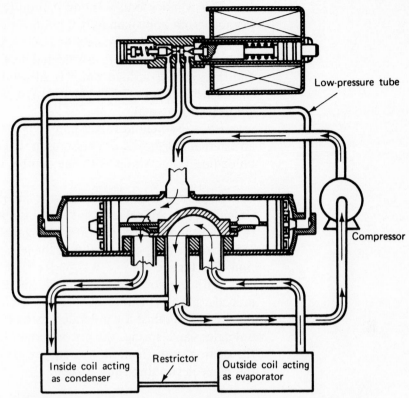

**Figure 5-7.** Energized solenoid coil. (Courtesy of Ranco Controls Division)

In the heating cycle, the solenoid coil is first energized to operate the pilot-valve plunger. See Figure 5-7. This action closes the left port with one needle valve and keeps the right port open with the other. The pressure differential, created between the two main valves and the chambers by pilot-valve action, instantly causes the two pistons to move the sliding post. Both end chambers soon arrive at equal pressures within an operating phase. However, this condition is instantly changed by action of the pilot valve in response to the temperature control action.

**COMPRESSOR-CAPACITY CONTROL**

Compressor-capacity control is usually in the form of compressor cylinder unloaders. They are available on all large capacity compressors. In most instances they are mounted on the compressor head of one or more cylinder banks, depending upon how much capacity control is needed. They are self-actuated, suction-pressure or electric-solenoid controlled and discharge-pressure operated. The electric solenoid types use a suction pressure control to sense the pressure and complete the electrical circuit to the solenoid coil. Valve operation is such that pressure of 25 pounds per square inch gauge (275.32 kilopascals) is established between the suction and discharge pressure.

Each control valve loads or unloads one compressor cylinder bank by allowing discharge gas to pass to the suction side of the cylinder through a bypass port. Unloaded cylinders operate with no pressure differential and, therefore, consume very little electric power.

The cylinder load point on suction-pressure-controlled valves is adjustable from approximately 0 pounds per square inch gauge (100.989 kilopascals) to approximately 85 pounds per square inch gauge (688.72 kilopascals). The pressure differential between the cylinder load-up point and the cylinder unload point is adjustable from about 5 pounds per square inch gauge (137.52 kilopascals) to 20 pounds per square inch gauge (240.87 kilopascals).

Electric solenoid valves unload controlled cylinders in response to an external thermostat or pressurestat. Pressure differential for complete unloading range varies with the control device used.

*Loaded Operation.* When the suction pressure is above the control point, the control set point spring is overcome. The diaphragm snaps to the left and relieves the pressure against the poppet valve. The drive spring moves the poppet valve to the left, and it seats in the closed position. See Figure 5-8.

With the poppet valve closed, the discharge gas is directed to the unloader piston chamber and pressure builds up against the piston. When the pressure against the unloader piston is high enough to overcome the unloader valve spring, the piston moves the valve to the right, opening the suction port. Suction gas can now be drawn into the cylinders, and the bank is running fully loaded.

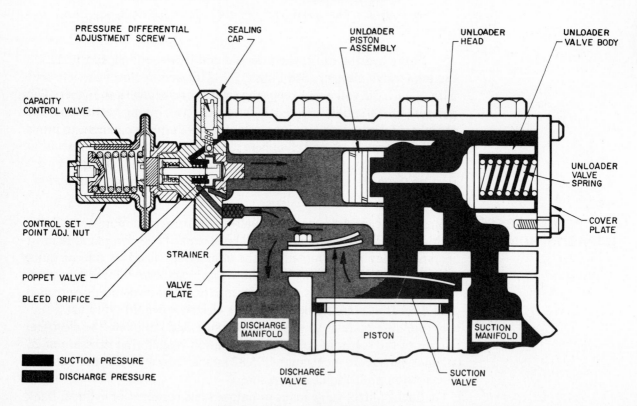

**Figure 5-8.** Loaded cylinder as used on Carrier compressor. (Courtesy of Carrier Corporation)

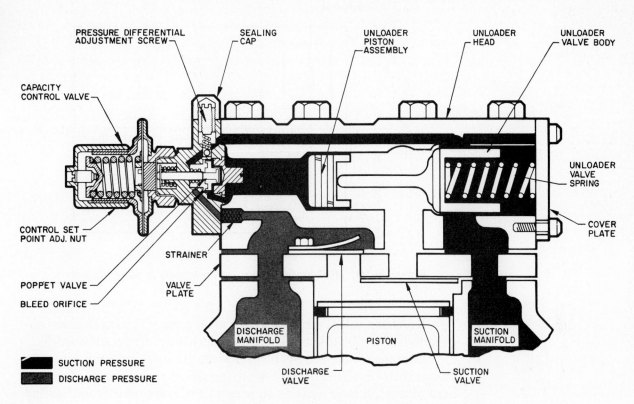

PRESSURE DIFFERENTIAL ADJUSTMENT SCREW

SEALING CAP

UNLOADER PISTON ASSEMBLY

UNLOADER HEAD

UNLOADER VALVE BODY

CAPACITY CONTROL VALVE

UNLOADER VALVE SPRING

COVER PLATE

CONTROL SET POINT ADJ. NUT

STRAINER

POPPET VALVE

VALVE PLATE

BLEED ORIFICE

DISCHARGE MANIFOLD

PISTON

SUCTION MANIFOLD

DISCHARGE VALVE

SUCTION VALVE

SUCTION PRESSURE

DISCHARGE PRESSURE

**Figure 5-9.** Unloaded cylinder as used on Carrier compressor. (Courtesy of Carrier Corporation)

*Unloaded Operation.* As the suction pressure falls below the control set point, the control spring expands snapping the diaphragm to the right. This forces the poppet valve open and allows gas from the discharge manifold to vent through the base of the control valve to the suction side. See Figure 5-9. Loss of full discharge pressure against the unloader spring allows the unloader valve spring to move to the left to close the piston. The suction port is blocked, isolating the cylinder bank from the suction manifold. The cylinder bank is now unloaded.

**COILS** Most solenoid coils are molded in an epoxy resin. The coil assembly is not waterproof, but it is moisture resistant. It will not usually break down electrically when high voltage is applied between the coil and the valve body after prolonged immersion in water at room temperature. The wire winding in these coils is like most other electromagnetic coils.

Solenoid valves may be used as a main liquid-line control valve on multiple systems as insurance against liquid floodback on compressor start-up. They may also be used to prevent liquid from entering the low side by leakage through the expansion valve during the off cycle. See Figure 5-10. Often solenoid valves are used to control the flow of refrigerant to individual evaporators, all of which are connected to the same compressor. These are only samples of uses of an indispensable control in the refrigeration industry.

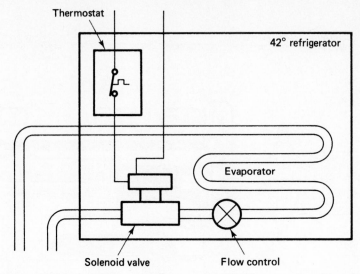

**Figure 5-10.**   Refrigerant flow controlled by a solenoid valve.

Reversing valves are used in two main areas: (1) in heat pump applications and (2) in hot-gas defrost systems in refrigeration work.

Compressor-capacity-control devices are used on compressors to change their capacities in response to the system requirements. They are used mostly on large systems to prevent the evaporator from frosting and to reduce power consumption during low-load conditions.

## REVIEW QUESTIONS

1. What is the operating force used in solenoid valves?
2. What devices are used on compressors to reduce power consumption?
3. What devices may be used on compressors to prevent evaporator frosting?
4. What device is used to divert refrigerant flow through a system?
5. Name the two types of solenoid valves.
6. On what types of systems are reversing valves used?
7. What is the most popular form of compressor capacity control?
8. In response to what do electric solenoid valves unload compressor cylinders?
9. What does the suction pressure do when a compressor unloads?
10. What is the purpose of compressor-capacity-control devices?

# 6

## Pressure Controls and Oil-Failure Controls

Any time that an electric motor is stalled or overloaded, the motor draws approximately six times its full-load amperage rating. If the overloaded condition is allowed to continue, the motor windings become overheated. The least that will happen is that the insulation on the motor windings will be destroyed, rendering the motor inoperative.

A compressor motor can be protected from overcurrent damage or bearing damage by use of a pressure control that will sense an excessively high or low refrigerant pressure. It can also be protected from bearing damage due to a lack of lubricating oil by the use of an oil-failure control.

**DEFINITIONS** The following are the accepted definitions of pressure controls and oil-failure controls.

**Pressure Controls:** *Pressure controls* are switching devices that are used to stop the compressor motor when the refrigerant pressures reach a predetermined point.

**Oil-failure Controls:** *Oil-failure controls* are switching devices that give dependable protection against major breakdowns on pressure-lubricated refrigeration compressors by guarding against low lubricating oil pressure.

*Operation.* In general, a low-pressure control is connected to the suction (low) side of the refrigeration system and is set to stop the compressor if the low-side refrigerant pressure drops to a predetermined level.

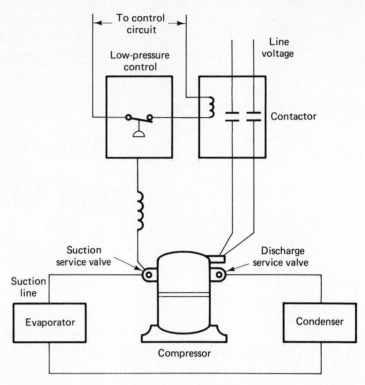

**Figure 6-1.**   Low-pressure control in refrigeration system.

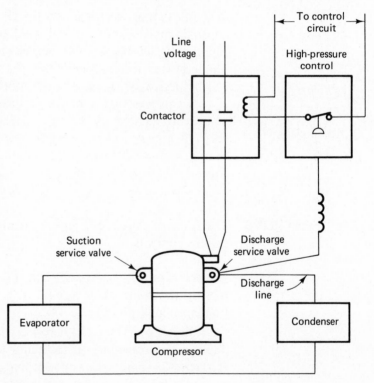

**Figure 6-2.**   High-pressure control connections in a refrigerant system.

See Figure 6-1. Suction- (low-) pressure controls are considered to be those that open on a fall in refrigerant pressure and close on a rise in refrigerant pressure.

High-pressure controls are generally connected to the high (discharge) side of the system and are set to stop the compressor if the high-side refrigerant pressure increases to a predetermined level. See Figure 6-2. High- (discharge-) pressure controls are considered to be those that open on a rise in refrigerant pressure and close on a fall in refrigerant pressure.

Both the high- and low-pressure controls are available in either automatic or manual reset models. The manual reset models are used to prevent operation of the equipment until it is checked out and the problem repaired before operation is continued to prevent damage to the compressor. They may be either adjustable or nonadjustable. See Figure 6-3.

(a)                    (b)      **Figure 6-3.** (a) Adjustable and (b) nonadjustable pressure controls. (Courtesy of Ranco Controls Division)

Pressure controls are available in either single-function controls or dual-function controls. The one chosen depends upon the equipment design and requirements and cost considerations. See Figure 6-4.

Pressure-control electrical ratings have a range of about 24 full-load amperes and 102 locked-rotor amperes on 120 volts or 240 volts.

The majority of uses for pressure controls are in the following areas:

Suction-pressure sensing for temperature control
Suction-pressure sensing for pumpdown control
Suction-pressure sensing for capacity control
Suction-pressure sensing for low-pressure limit control

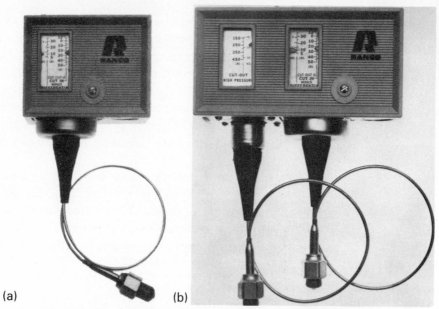

(a)                    (b)

**Figure 6-4.** (a) Single-function and (b) dual-function pressure controls. (Courtesy of Ranco Controls Division)

Suction-pressure sensing for alarm control

High-pressure sensing for high-pressure control

High-pressure sensing for condenser-fan control

### Suction-Pressure Sensing for Temperature Control

*Suction-pressure sensing devices* open on a fall in pressure and close on a rise in pressure. They are normally used on thermostatic expansion-valve systems and control the temperature by sensing the suction pressure and the resulting temperature of the evaporator. The compressor is stopped in response to the suction pressure, which is directly related to the evaporator temperature. See Figure 6-5.

The low-pressure control is adjusted to start (cut in) the compressor motor when the suction pressure rises to a predetermined maximum point, which is determined by the temperature of the product that is stored in the space being cooled. It will then stop (cut out) the compressor motor when the suction pressure has fallen to a predetermined minimum point. See Table 6-1.

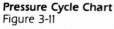
**Pressure Cycle Chart**
Figure 3-11

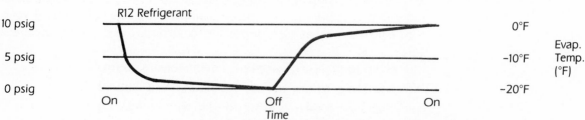

**Figure 6-5.** Suction pressure sensing for temperature cycle chart (Courtesy of Ranco Controls Division).

**TABLE 6-1**
Approximate Pressure Control Settings.

| | VACUUM: ITALIC FIGURES | | GAGE PRESSURE: LIGHT FIGURES | | | |
|---|---|---|---|---|---|---|
| | | | Refrigerant | | | |
| | 12 | | 22 | | 502 | |
| Application | Out | In | Out | In | Out | In |
| Ice cube maker, dry type coil | 4 | 17 | 16 | 37 | 22 | 45 |
| Sweet water bath, soda fountain | 21 | 29 | 43 | 56 | 52 | 66 |
| Beer, water, milk cooler, wet type | 19 | 29 | 40 | 56 | 48 | 66 |
| Ice cream, hardening rooms | 2 | 15 | 13 | 34 | 18 | 41 |
| Eutectic plates, ice cream truck | 1 | 4 | 11 | 16 | 16 | 22 |
| Walk-in, defrost cycle | 14 | 34 | 32 | 64 | 40 | 75 |
| Vegetable display, defrost cycle | 13 | 35 | 30 | 66 | 38 | 77 |
| Vegetable display case, open type | 16 | 42 | 35 | 77 | 44 | 89 |
| Beverage cooler, blower, dry type | 15 | 34 | 34 | 64 | 42 | 75 |
| Retail florist, blower coil | 28 | 42 | 55 | 77 | 65 | 89 |
| Meat display case, defrost cycle | 17 | 35 | 37 | 66 | 45 | 77 |
| Meat display case, open type | 11 | 27 | 27 | 53 | 35 | 63 |
| Dairy case, open type | 10 | 35 | 26 | 66 | 33 | 77 |
| Frozen food, open type | -7 | 5 | 4 | 17 | 8 | 24 |
| Frozen food, open type, thermostat | 2°F | 10°F | — | — | — | — |
| Frozen food, closed type | 1 | 8 | 11 | 22 | 16 | 29 |

The distance between the cut-in and cut-out pressures is known as the *differential*. When the control is adjusted for a narrow differential, the temperature variations will also be narrow. However, a narrow differential may cause short-cycling of the compressor. A wide differential allows the compressor to operate with longer on-cycles, but there may also be wide temperature variations inside the refrigerated area.

**Suction-Pressure Sensing for Pumpdown Control**

Suction-pressure sensing devices open on a fall in pressure and close on a rise in pressure. This type of control is generally used on systems in which the evaporator is located a long way from the condensing unit and/or on systems in which the condensing unit is located outside the building.

The advantage of using a pumpdown control is to remove the refrigerant from the low side of the system. This prevents liquid slugging of the compressor and a possible loss of the lubricating oil on the next start-up of the system. In these types of systems, a temperature control energizes and de-energizes a liquid-line solenoid valve causing the system either to stop or to operate. See Figure 6-6.

When the space requires cooling, the temperature control energizes the liquid-line solenoid and permits refrigerant to flow into the evaporator. As a result, the low-side pressure rises to the cut-in setting of the low-pressure control. The low-pressure control then starts the compressor motor. When the space has cooled down and the temperature control is satisfied, the solenoid valve is de-energized, stopping the flow

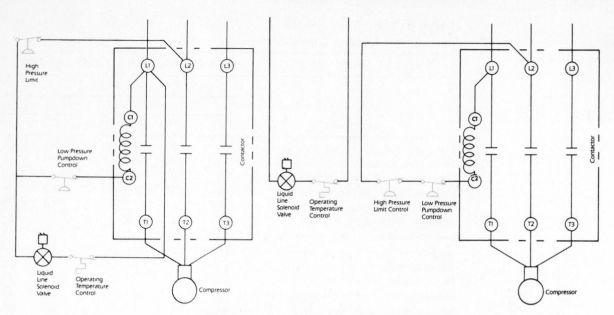

**Figure 6-6.** Suction pressure sensing for pumpdown-control connections. (Courtesy of Ranco Controls Division)

of refrigerant to the evaporator. The compressor continues to run until the refrigerant has been pumped from the low side of the system and the pressure falls to the cut-out setting of the low-pressure control. The compressor motor is then stopped.

The settings used on controls for pumpdown control are very important. It is the cut-in setting that determines how high the refrigerant pressure in the low side of the system must go before the compressor motor starts. Therefore, it must be carefully selected. The equipment manufacturer generally makes recommendations for this setting.

The cut-in setting for condensing units located outside is determined by selecting either the coldest unit operating temperature or the coldest anticipated outdoor ambient temperature, whichever is colder. The equipment manufacturer generally makes recommendations for this setting.

**Suction-Pressure Sensing for Capacity Control**

Suction-pressure sensing controls are designed to open on a rise in pressure and close on a fall in pressure. They are used to energize and de-energize electric solenoid-controlled compressor-capacity-control devices. As the load decreases (the suction pressure falls) the unloader solenoid is de-energized, causing a reduced compressor capacity. There is generally one suction-pressure-sensing control used for each compressor unloader. See Figure 6-7.

Settings for pressure controls used as unloader controls are higher than the low-pressure control settings for other uses. They are set to maintain a desired evaporator temperature which is generally recommended by the equipment manufacturer.

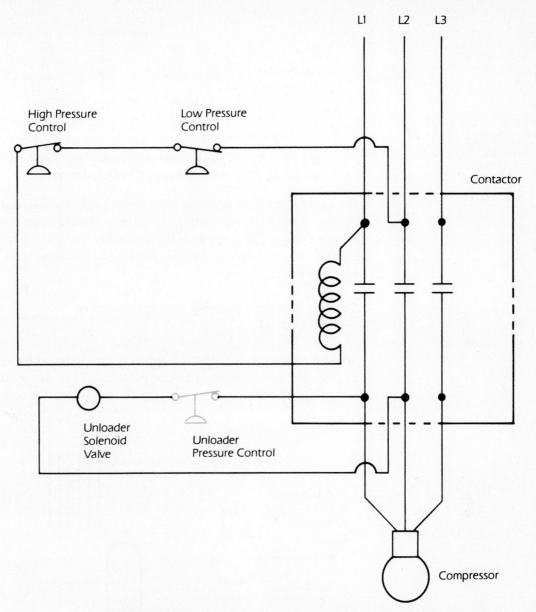

**Figure 6-7.** Suction pressure sensing for capacity-control connections. (Courtesy of Ranco Controls Division)

### Suction-Pressure Sensing for Alarm Control

These suction-pressure-sensing controls are designed to close on a rise in pressure and open on a fall in pressure. These are most popular on systems with remote evaporators. An abnormally high temperature is indicated by a high-suction pressure, which closes the contacts in the pressure control and energizes the alarm control. See Figure 6-8.

Generally, the low-pressure control is used in conjunction with an automatic-reset time-delay relay. The low-pressure control energizes the time-delay relay and if the suction pressure remains higher than desired

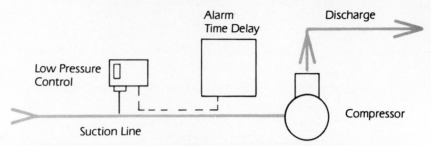

**Figure 6-8.** Suction pressure sensing for alarm-control connections. (Courtesy of Ranco Controls Division)

for a given period of time, the relay contacts close and energize the alarm device, alerting the personnel that the temperature is too high in the conditioned space. The purpose of the time delay is to prevent nuisance alarm signals. When a defrost cycle is used, the time delay may be from 45 to 60 minutes.

### Suction-Pressure Sensing for Low-Pressure Limit Control

These suction-pressure-sensing controls are designed to open on a fall in pressure and close on a rise in pressure. They are used on systems not equipped with a pumpdown-type control to provide compressor pro-

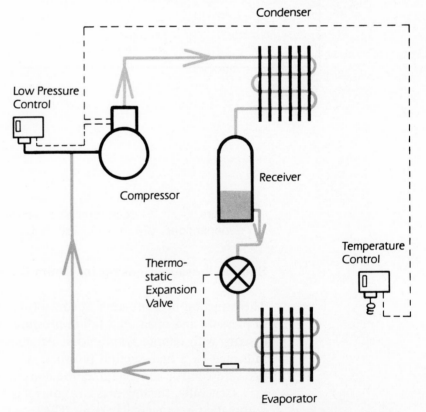

**Figure 6-9.** Suction pressure sensing for low-pressure control connections. (Courtesy of Ranco Controls Division)

tection when a low suction pressure occurs, indicating a possible loss of refrigerant charge or an abnormally low evaporator temperature. They are sometimes used on air conditioning systems to prevent frosting of the evaporator and to prevent freezing of the evaporator in a water-chiller-type system. When the suction pressure reaches a predetermined point, these controls stop the compressor to prevent damage to the evaporator. See Figure 6-9.

In most cases the manual reset control is preferred for low-pressure protection. However, when perishable products are being stored, an automatic reset type may be used to prevent loss of the product when the problem is only a momentary one.

Another type control is to include a manual-reset time-delay relay in conjunction with an automatic-reset low-pressure control having contacts that open on a rise in pressure and close on a fall in pressure. This system almost eliminates system shutdown due to momentary problems that sometimes exist while at the same time providing the equipment protection offered by use of a manual reset control.

In this type of system, when the control contacts close, the time-delay device is energized. If the suction pressure does not increase to the open setting of the control contacts, the time delay will shut the system down. The time-delay device must be reset before system operation will be resumed. Typically, a 2- to 5-minute time delay is used. See Figure 6-10.

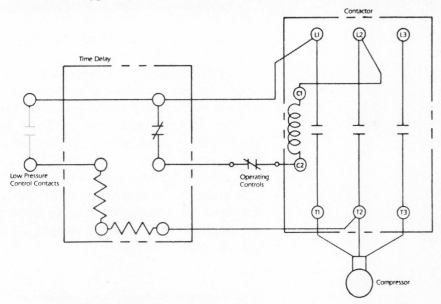

**Figure 6-10.** Suction pressure limit with time delay. (Courtesy of Ranco Controls Division)

## High-Pressure Sensing for High-Limit Control

These high-pressure-sensing controls are equipped with contacts that open on a rise in pressure and close on a fall in pressure. High-pressure controls are used to sense the discharge pressure and stop the compressor when an abnormally high discharge pressure occurs. See Figure 6-11.

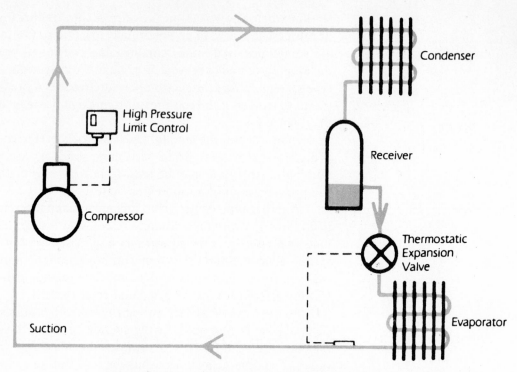

**Figure 6-11.** High-pressure sensing for high-limit control connections. (Courtesy of Ranco Controls Division)

The pressure-control settings vary with the different types of refrigerants and different system designs. The equipment manufacturer generally recommends the proper settings for the equipment under consideration. Most manufacturers prefer that manual-reset high-pressure controls be used to provide protection for their equipment.

In some installations, an SPDT-type switching arrangement is used. The NO contacts are used to energize an alarm circuit, indicating that the switch contacts have opened and resetting is required. The NC contacts control the operation of the compressor motor.

When the high-pressure control opens, there is a problem with the equipment that must be corrected, or damage to the compressor will probably occur. This is the reason that the manual-reset control is preferred over the automatic-reset type.

### High-Pressure Sensing for Condenser Fan Control

These high-pressure-sensing controls are equipped with contacts that close on a rise in pressure and open on a fall in pressure. They are used on air-cooled systems where it is desirable to maintain the discharge pressure at a certain point in low ambient temperatures. The higher pressure is required to assure that proper refrigerant is being fed into the evaporator for proper system operation.

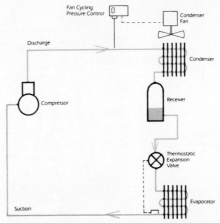

**Figure 6-12.** High-pressure sensing for condenser fan control. (Courtesy of Ranco Controls Division)

In this type of application the switch contacts close when the discharge pressure reaches a predetermined point to start the condenser fan motor, and they open when the discharge pressure has fallen to a predetermined point to stop the condenser fan. This keeps the discharge pressure up. See Figure 6-12.

These controls should be adjusted correctly for proper operation. Too wide a differential will allow the fan motor to remain off for too long a time, resulting in wide variations in the discharge pressure and improper feeding of the thermostatic expansion valve. A differential that is too narrow will cycle the fan too rapidly, causing a shortened fan-motor life. The equipment manufacturer will generally make recommendations for these control settings.

When more than one fan is to be cycled, a separate control is used for each fan motor. In most cases multiple controls are set with about 10 pounds per square inch guage (171.97 kilopascals) difference between them. This difference allows for fan staging and better control of the discharge pressure.

## PROPORTIONAL PRESSURE CONTROLS

*Proportional pressure controls* are designed to control motor actuators for the control of dampers, valves, and other similar devices. Some models have a throttling adjustment to permit on-the-job selection of the correct throttling range for stable system control. See Figure 6-13. These controls may be used along with other necessary aids to control the flow of water through a water-cooled condenser, to position face and bypass dampers on multizone systems, to control the flow of water through a chilled- or hot-water coil, or to control the position of outdoor air dampers for energy conservation. They should be used with the manufacturer's recommended accessory equipment for proper system control.

*Operation.* These controls operate the same as modutrol motors described in Chapter 14.

**Series P80**

ACTION ON INCREASE
OF PRESSURE

**Figure 6-13.** Proportional pressure control. (Courtesy of Johnson Controls, Inc., Control Products Division)

## OIL-FAILURE CONTROLS

*Oil-failure controls* are designed to provide compressor protection by guarding against low lubrication-oil pressure. See Figure 6-14. The oil-failure control is a differential pressure control, which was developed to monitor the effective oil pressure and stop the compressor in case of oil failure.

These controls have two opposing pressure-sensing elements that sense the *net* oil pressure. Net oil pressure is the difference between the crankcase pressure and the oil-pump outlet pressure. The compressor crankcase is usually operating at a pressure other than atmospheric. Therefore, it is necessary to measure this net pressure to determine whether or not proper lubrication is available. These pressure-sensing elements operate a set of NC contacts that open on an increase in pressure difference.

A built-in time-delay switch allows for oil-pressure pickup on starting and avoids nuisance shutdowns on oil-pressure drops of short duration during the running cycle. See Figure 6-15.

*Operation.* In operation the total oil pressure is the combination of the crankcase pressure and the pressure generated by the oil pump. The net oil pressure available to circulate oil is the difference between the total oil pressure and the refrigerant pressure in the crankcase: Total oil pressure minus refrigerant pressure equals net oil pressure. This control measures that difference in pressure, or net oil pressure.

When the compressor starts, a time-delay switch is energized. If the net oil pressure does not increase to the control cut-in setting within the required time limit, the time-delay switch trips to stop the compressor. If the net oil pressure rises to the cut-in point within the required time after the compressor starts, the time-delay switch is automatically de-energized and the compressor continues to operate normally. If the net oil pressure should fall below the cut-out setting during the running cy-

**Figure 6-14.** Oil-failure control. (Courtesy of Ranco Controls Division)

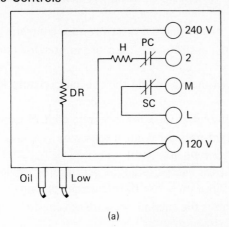

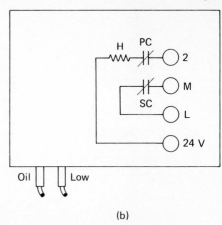

(a)                                                                            (b)

**Figure 6-15.** Internal diagram for Penn line voltage and low-voltage controls. (Courtesy of Control Products Division, Johnson Controls, Inc.)

cle, the time-delay switch is energized, and unless the net oil pressure returns to the cut-in point within the time delay period, the compressor stops.

The time-delay switch is a trip-free thermal-expansion device. It is compensated to minimize the affect of ambient temperatures from 32 °F (0 °C) to 150 °F (66 °C). The timing is also affected by voltage variations. The manufacturer's recommendations should be followed during the installation and adjustment of these controls. Normally, connect the oil-pressure line to the pressure connection labeled OIL and the crankcase line to the pressure connection labeled LOW. Wire as suggested for the specific equipment under consideration. See Figure 6-16. Oil-failure controls are electrically rated for pilot duty only.

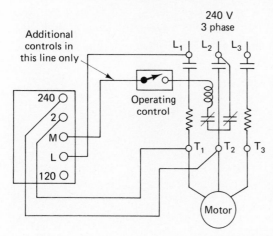

**Figure 6-16.** Penn control wiring diagram used on 230-volt system. (Courtesy of Control Products Division, Johnson Controls, Inc.)

## REVIEW QUESTIONS

1. In general, what refrigerant pressure are low-pressure controls designed to sense?

2. What type of pressure control is designed to sense the discharge pressure in a refrigeration system?

3. What is the purpose of the manual-reset pressure control?

4. What type of pressure control opens on a rise in pressure and closes on a fall in pressure?

5. The difference between the cut-in and the cut-out pressures is known as what?

6. What will a wide differential on a low-pressure control cause?

7. What type of refrigerant-pressure control is used on systems in which the evaporator is located a long way from the condensing unit?

8. Are the controls used for suction-pressure sensing for capacity control designed to open or close on a rise in refrigerant pressure?

9. In what type of system is the suction-pressure sensing for alarm control most popular?

10. What control is used in conjunction with the low-pressure control in the suction-pressure sensing for alarm control?

11. Are the contacts in the high-pressure sensing for high-limit control designed to open or close on a rise in refrigerant pressure?

12. Is the manual- or automatic-reset high-pressure control preferred?

13. On what type of systems are high-pressure-sensing controls for the condenser fan used?

14. What will be the result of too wide a differential on a high-pressure-sensing control for the condenser fan?

15. How do oil-failure controls provide compressor protection?

# 7 Temperature Controls, Humidistats, and Airstats

Any building, regardless of location, age, or architecture, can now provide its owner with year-round comfort and safety with a well-designed, properly installed, and carefully controlled air conditioning system. Maintaining perishable foods at the necessary temperature would also be very difficult, at best, if it were not for the development of modern temperature controls.

**DEFINITIONS** The following are the accepted definitions of temperature controls, humidistats, and airstats.

**Temperature Controls:** *Temperature controls* are temperature-sensing devices used to maintain a desired temperature by switching on or off the necessary equipment in response to space demands.

**Humidistats:** *Humidistats* control the humidity within a structure by controlling the operation of a humidifier.

**Airstats:** *Airstats* are safety devices used to control ventilation fans during abnormal conditions.

**TEMPERATURE CONTROLS** Temperature controls are generally divided into two different types: (1) refrigeration temperature control and (2) air conditioning temperature control.

**Refrigeration Temperature Controls**

*Refrigeration temperature controls* are used to signal the equipment when to operate and when not to operate in response to the needs of the space being cooled, whether it is a commercial refrigeration case or a domestic refrigerator.

In a commercial case the temperature control is used to actuate a motor contactor to either start or stop the compressor on temperature demand. See Figure 7-1.

Refrigeration temperature controls are manufactured with the temperature-sensing device contained within temperature-sensitive power elements. See Figure 7-2.

The vaporization or vapor-pressure element is based on the principle that the boiling temperature of a liquid depends upon the vapor pressure at the liquid surface. By partially filling a bulb with liquid and

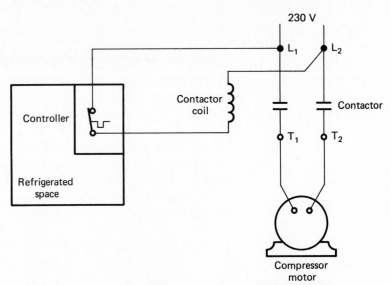

**Figure 7-1.** Wiring diagram for refrigeration temperature control.

**Figure 7-2.** Liquid-filled remote bulb controller.

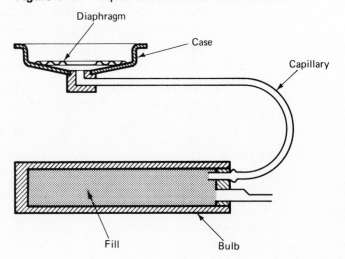

**Figure 7-3.** Refrigeration thremostat. (Courtesy of Control Products Division, Johnson Controls, Inc.)

connecting the vapor space to a pressure sensitive element, a closed system is formed in which the vapor pressure will depend upon the temperature of the bulb. In this type of element the pressure-sensitive mechanism is usually a bellows. See Figure 7-3.

### Methods of Refrigeration Temperature Control:

The most popular methods of refrigeration temperature control are (1) space and return air sensing for temperature control, (2) evaporator temperature sensing for temperature control, (3) product temperature sensing for temperature control, and (4) ice thickness sensing for temperature control.

*Space and Return-Air Sensing for Temperature Control.* This method of temperature control is used on most walk-in refrigeration applications. The contacts in this control close on a rise in temperature and open on a fall in temperature of the refrigerated space. See Figure 7-3. Some thermostats used for this purpose have an NO set of contacts that close when the space temperature has risen a few degrees above the cut-in point for normal operating temperatures. The NO contacts, when closed, energize an alarm circuit that notifies the user that the space temperature is higher than desired.

The major advantage of this method of control is the close control of the refrigerated space temperature. Temperature controls that sense the space temperature by use of air coils are popular on systems such as walk-in coolers and freezers. See Figure 7-4.

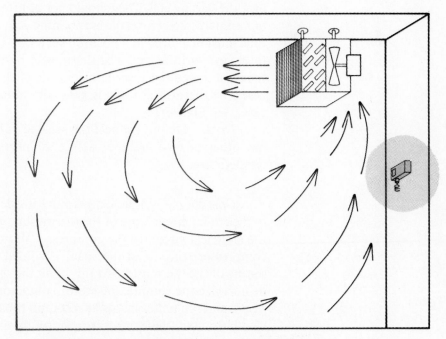

**Figure 7-4.** Location of refrigeration thermostat for a walk-in cooler. (Courtesy of Ranco Controls Division)

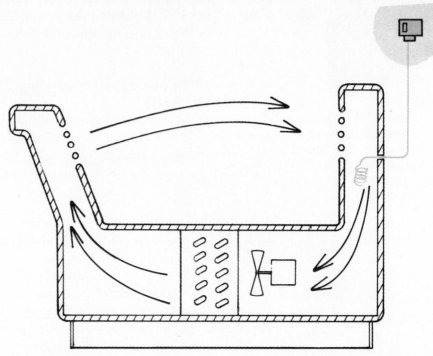

**Figure 7-5.**   Thermostat location on a display case. (Courtesy of Ranco Controls Division)

These controls should be installed in a place where they will not be easily damaged or where the product will prevent a free air circulation over the sensing coil. They should not be affected by the opening of doors or air flow directly from the evaporating coil. The sensing bulb or capillary tube must be placed in a position such that it will sense the air returning to the evaporating coil. The thermostat body may be installed either in the conditioned space or on the outside of the case. When used in display cases the thermostat body is generally mounted on a wall close to the case. See Figure 7-5.

*Note:*   Do not connect the sensing bulb to the evaporating coil or any other surface that may affect the proper sensing of the return air temperature.

*Operation.*   As the temperature inside the conditioned space rises to the cut-in temperature of the thermostat, its contacts close, completing the electrical circuit to the compressor motor contactor or starter. The compressor starts, and the space is cooled to the cut-out temperature setting of the thermostat. At this point the thermostat contacts open and de-energize the compressor motor contactor or starter, the compressor stops, and the unit remains at rest until the temperature has risen sufficiently to start the compressor motor again.

**Figure 7-6.** Thermostat used for product-temperature sensing. (Courtesy of White-Rodgers Division, Emerson Electric Co.)

If an alarm signal is included, it remains unused until the temperature rises a given number of degrees above the cut-in setting of the thermostat. Then the alarm contacts close and either start a bell ringing or a light burning, whichever is used, to alert the user that service for the equipment is required.

*Product-temperature Sensing for Temperature Control.* Product-temperature sensing is a popular control method used on liquid chillers and soft ice cream, slush ice, and yogurt machines. The contacts in these types of controls close on a rise in temperature and open on a fall in temperature of the product being refrigerated. They provide close temperature control by directly sensing the temperature of the product. See Figure 7-6. These controls generally have a narrow differential to aid in providing the required close temperature control of the product.

When sensing the temperature of a liquid, such as wine or a liquid in a bulk milk tank, the control element is usually inserted into a well that is surrounded by the product being refrigerated. The well is securely fastened to the tank and is insulated from the outside air temperature. See Figure 7-7.

When a soft ice cream, slush drink, or yogurt machine is being used, the sensing element either may be installed in a well, if provided, or it may simply be wrapped around and secured to the freezing drum. See Figure 7-8.

*Operation.* In operation, if the product temperature rises to the cut-in setting of the thermostat, the electrical circuit is completed to the compressor motor contactor or starter. The compressor motor is then started and the product is cooled down to the desired temperature. The thermostat contacts then open, interrupting the electrical circuit to the contactor or starter-holding coil, and the compressor motor is stopped. The product temperature then begins to rise again and the cycle is repeated.

**Figure 7-7.** Thermostat installation in a bulk milk tank. (Courtesy of Ranco Controls Division)

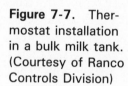

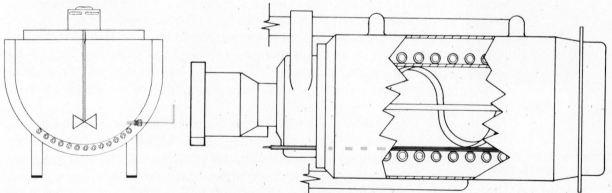

**Figure 7-8.** Thermostat installation in a soft ice cream, slush drink machine, or yogurt machine. (Courtesy of Ranco Controls Division)

*Evaporator-Temperature Sensing for Temperature Control.* Evaporator-temperature sensing to control temperature is popular on medium- and low-temperature self-contained refrigeration systems using a capillary tube as the flow-control device. This method of control also assures that the evaporator is completely defrosted during the off cycle, because the evaporator will not reach the cut-in setting until the evaporator coil is defrosted. The contacts close on a rise in temperature and open on a fall in temperature. The temperature of the evaporator can be sensed with either a capillary tube or a capillary tube and bulb. The sensing element may be installed in many ways to sense the evaporator temperature—on a return bend of the coil, on a plate that is secured to several return bends, wrapped around some tubes in the evaporator, or inserted into a well provided for this purpose. If it is inserted into a well, a sealant must be used to prevent moisture from entering the well. At least 6 inches of the capillary tube must be in direct contact with the surface being sensed so that proper control is maintained.

Components used for this method of control are equipped with a wide differential. When the proper differential is selected and used, the compressor will not short cycle.

*Operation.* When the surface temperature of the evaporator has risen sufficiently, the contacts of the thermostat will close and complete the electrical circuit to the compressor motor contactor or starter-holding coil and start the compressor motor. The evaporator surface temperature falls until the cut-out setting of the thermostat is reached. Then the contacts open, de-energizing the compressor contactor or starter-holding coil, and the compressor stops. The evaporator surface temperature begins to rise to the cut-in setting of the thermostat and the cycle is started again.

*Ice-Thickness Sensing for Temperature Control.* Ice-thickness sensing for temperature control is popular on ice-bank installations. The contacts in these controls close on a rise in temperature and open on a fall in temperature of the water in the tank. The ice-bank control is designed to control the refrigeration compressor so that the desired volume of ice is built up on the evaporator coil. See Figure 7-9. The bulb of this control is filled with water. The sensing bulb is mounted close to the evaporator coil so that it can sense the buildup of ice. See Figure 7-10.

*Operation.* When the temperature of the water in the ice storage tank warms up to the cut-in setting of the thermostat, the electrical circuit to the compressor motor contactor or starter-holding coil is completed and the contactor or starter is energized, starting the compressor motor.

When the water temperature has fallen to the cut-out setting of the thermostat, the water in the thermostat bulb also freezes and expands. This expansion is transmitted to the control through the liquid filled capillary to open the thermostat contacts and de-energize the holding coil circuit, stopping the compressor motor.

**Figure 7-9.** Thermostat used on ice bank systems. (Courtesy of Ranco Controls Division)

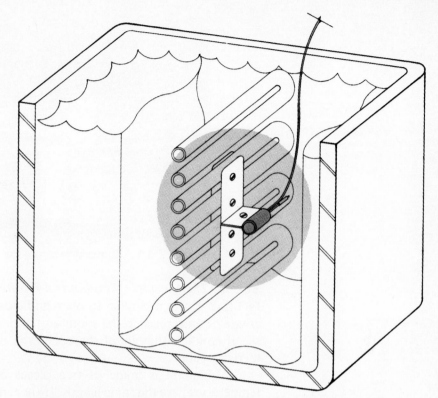

**Figure 7-10.** Thermostat installation in an ice bank. (Courtesy of Ranco Controls Division)

### Air Conditioning (Room) Thermostats

A temperature control with which we are more familiar is generally called a *room thermostat*. A room thermostat is used for sensing the room temperature and signaling the air conditioning equipment either to operate or to stop operating in response to this temperature.

In its simplest form, a thermostat is a device that responds to changes in air temperature and causes a set of electrical contacts to open or close. This is the basic function of a thermostat, but there are many different types, designed to perform a variety of switching functions. They are available in heating, cooling, or heating and cooling types.

One of the early types of heating systems that was capable of some degree of automatic control was the hand-fired coal furnace. Thermostatic control of this system was accomplished with a SPDT thermostat and damper motor. This was a long way from the completely automatic systems of today, but it was the beginning of automatic control for residential and commercial building heating systems.

*Types of Room Thermostats.* There are three types of electric thermostats used in heating and cooling systems today: (1) the bimetal type, (2) the bellows-actuated type, and (3) the proportioning (modulating) type. The bimetal type is by far the most popular type in use today.

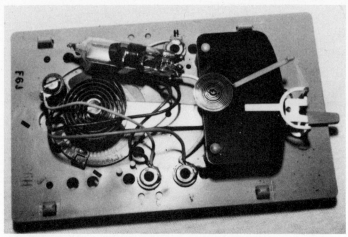

**Figure 7-11.** Bimetal thermostat with cover removed.

*Bimetal Thermostat:* The *bimetal thermostat* gets its name from the fact that it uses a bimetal to open and close a set of contacts on an increase or a decrease in the room air temperature. See Figure 7-11. This set of contacts may be of the open type or they may be enclosed in a mercury tube.

A bimetal is made of two pieces of metal, which, at a given temperature, are the same length. If the temperature of these two pieces of metal is increased, one becomes longer than the other because they are different types of metal, having different rates of expansion. These two metals are welded together in such a way that they become one solid piece, but they still keep their individual characteristics of different rates of expansion. See Figure 7-12.

When heat is applied, one piece expands at a faster rate than the other piece. In order for one piece to become longer than the other, it must bend the entire bimetal into an arc. See Figure 7-13.

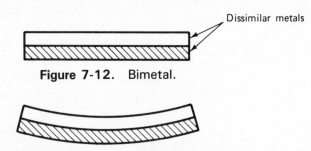

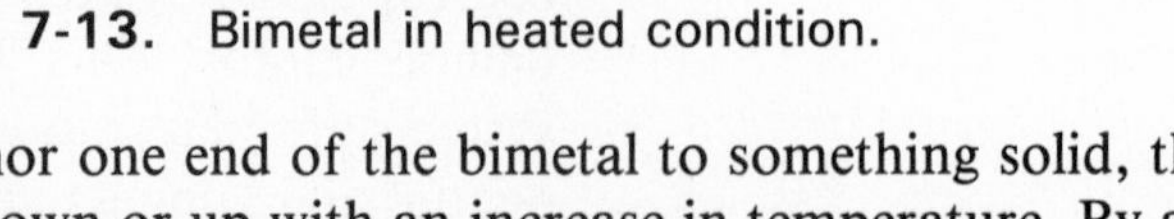

Dissimilar metals

**Figure 7-12.** Bimetal.

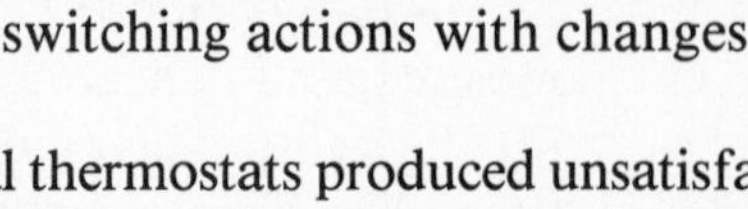

**Figure 7-13.** Bimetal in heated condition.

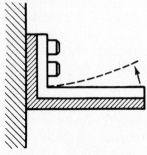

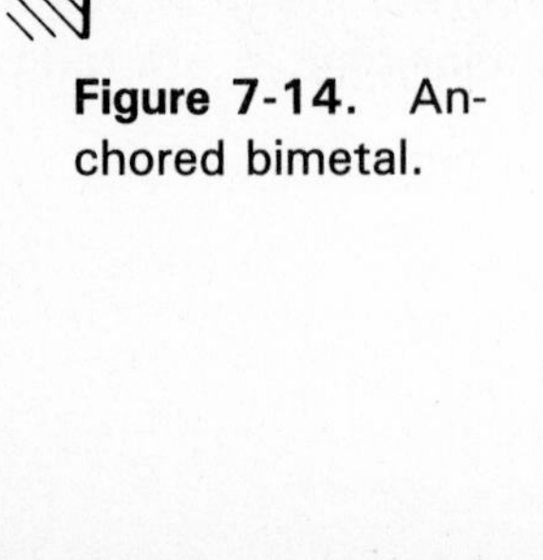

**Figure 7-14.** Anchored bimetal.

If we now anchor one end of the bimetal to something solid, the free end will move down or up with an increase in temperature. By attaching contacts to the free end and placing a stationary contact nearby, we can get different switching actions with changes in temperature. See Figure 7-14.

The first bimetal thermostats produced unsatisfactory results because of unstable action of the contacts. Due to the relatively small differences in room air temperature, the bimetal could not develop enough contact

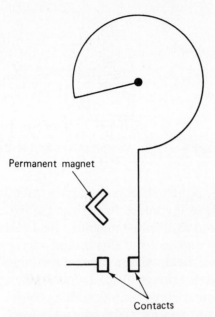

**Figure 7-15.** Thermostat-type bimetal.

pressure to obtain a positive electrical connection. With the development of the permanent magnet, it was possible to obtain the convenience of a control system incorporating the best features of modern control circuits. See Figure 7-15.

*Snap Action Versus Mercury Switch:* Room thermostats are available with either snap action or mercury switches. *Snap-action switches* are constructed with a fixed contact securely attached to the base of the thermostat. This contact is mounted inside a round permanent magnet, which produces a magnetic field in the area of the contact. The movable contact is attached to the bimetal and upon a decrease in temperature (on heating models) moves slowly toward the fixed contact. See Figure 7-16.

As the movable contact enters the magnetic field around the fixed contact, the magnetic field pulls the movable contact against the fixed contact with a positive snap. Because the movable contact has a floating action, it closes with a clean snap, that is, without contact bounce. This floating action also eliminates any tendency for the contacts to "walk" while opening. Either the walking action or a lack of the positive snap will cause arcing between the contacts, which in time will burn and pit them, reducing the electrical continuity between them. This will eventually cause the thermostat to become less responsive to temperature changes or not to make a circuit through the contacts at all.

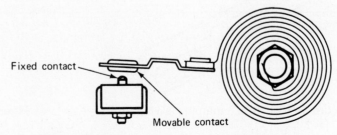

**Figure 7-16.** Typical thermostat bimetal.

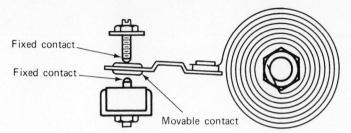

Fixed contact

Fixed contact

Movable contact

**Figure 7-17.**   Single-pole–double-throw thermostat.

As the bimetal becomes warmer (on a heating model), it wants to pull the movable contact away from the fixed contact. But, because the movable contact is in the magnetic field surrounding the fixed contact, the bimetal does not—at this instant—have enough force to overcome the magnetic field. As the bimetal continues to warm up and bend, it soon develops enough force to overcome the magnetic field and the movable contact breaks away with a positive snap. When we use SPDT switch action, another fixed contact is used. This contact is located so that when one contact is broken, another is made. See Figure 7-17.

All snap-action thermostats are supplied with dust covers to prevent dirt and other contaminants from getting on the contacts. Should it become necessary to clean these contacts, never use a file or sandpaper. A clean business card or smooth cardboard should be inserted between the contacts. With gentle pressure on the movable contact, pull the card back and forth to clean the dirt or film from the contacts.

*Mercury switches* perform the same switching action as snap-action switches, but the switching action is accomplished by a globule of mercury moving between two or three fixed probes sealed inside a glass tube. Two probes are used on SPST switches. See Figure 7-18. The SPDT models have three probes. See Figure 7-19.

These mercury switches are attached to the thermostat bimetal and perform the desired function.

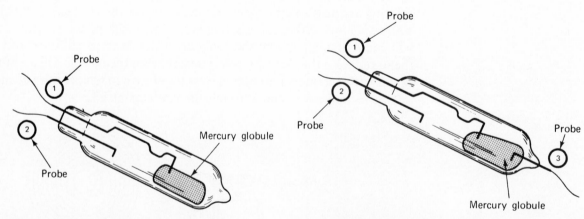

Probe

Probe

Mercury globule

**Figure 7-18.**   SPST mercury switch.

Probe

Probe

Probe

Mercury globule

**Figure 7-19.**   SPDT mercury switch.

### Thermostat Anticipators

There are two types of thermostat anticipators: (1) heating anticipator and (2) cooling anticipator. Their names are descriptive of their uses. These devices add artificial heat to the inside of the thermostat to aid in temperature regulation of the conditioned space. Without them, the temperature swings would be quite wide and would not be comfortable to most of us. The following is a description of the anticipators and how they affect the overall system operation.

*Heating Anticipators.*   *Heat anticipators* are placed in the thermostat circuit in series with the control circuit so that all current flowing in the control circuit must pass through the anticipator. The anticipator produces heat in an amount equal to the amount of current passing through it. This heat is released inside the thermostat and causes it to cycle more often than it would without anticipation. See Figure 7-20.

Heat anticipators are made in two types, fixed and adjustable. A fixed heat anticipator can be either a wire-wound or a carbon-resistor type. Earlier models used the wire-wound fixed anticipators. They were dipped in an insulating material and color-coded to indicate the primary control with which they were to be used.

Present-day thermostats with fixed anticipators use tubular resistors, that are color-coded to indicate the current draw of the primary control. These are supplied either as *nonremovable* (riveted in place) or *removable* (attached with a screw). Fixed anticipators must match the current draw of the control circuit. A number of different ranges are available. See Figure 7-21.

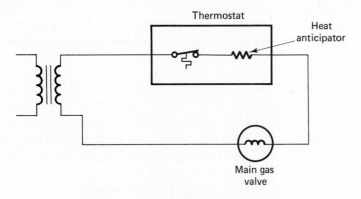

**Figure 7-20.**   Heating anticipator.

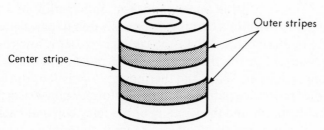

**Figure 7-21.**   Fixed heat anticipator.

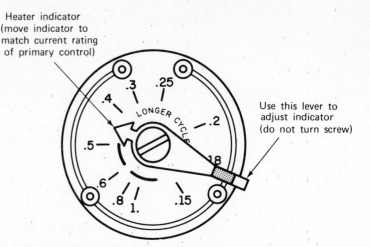

**Figure 7-22.** Adjustable heat anticipator.

The most versatile heat anticipator is, of course, the adjustable type. See Figure 7-22.

*Setting An Adjustable Heat Anticipator:* The primary purpose of an adjustable anticipator thermostat is to provide a single thermostat to match almost any type of situation encountered in the field. Before installing the thermostat, check the amperage draw of the temperature-control circuit to which it is to be connected. Make certain that the voltage of the control circuit and that of the thermostat are the same. Otherwise the thermostat could be damaged or the system may not operate properly. After making certain that the voltages are the same and the amperage draw has been determined, set the heat anticipator.

For example, if the amperage draw has been determined to be 0.45 amps, set the indicator on the adjustable anticipator to 0.45. See Figure 7-22. By matching the adjustable anticipator to the current draw of the control circuit, you are assured of the best possible heat anticipation for the system.

*Cooling Anticipators.* The *cooling anticipator* is a 0.25-watt carbon-type resistor similar to that found in a radio or television set. It is nonadjustable. This type of anticipation is known as *off-cycle anticipation.* To understand better the operation of this type of anticipation, look first at how it is used in a typical cooling system. See Figure 7-23.

In a cooling system, heat is added to the thermostat bimetal during the off cyle. This is just the opposite of heat anticipation in a heating system. The anticipator in a cooling system is wired in parallel with the contacts of the thermostat. When the contacts close, a low resistance path for the current in the thermostat circuit pulls in the relay on the air conditioning system. When the thermostat is satisfied, the contacts open. A higher resistance path from the transformer now exists through the cooling anticipator and the winding of the relay coil and back to the transformer. Because there is a high resistance in the cooling anticipator, the voltage drops to the point that the relay will not pull in.

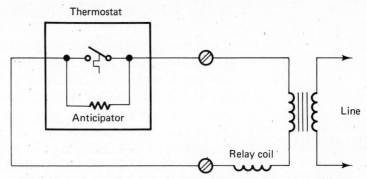

**Figure 7-23.** Off-cycle anticipation schematic diagram.

The current flowing through the cooling anticipator heats up the thermostat bimetal. This causes the bimetal to be warmer than the surrounding room air temperature, which is also rising (the system is off). This false heat causes the contacts to close before the room air temperature reaches the cut-in point. Thus we are bringing the cooling system on sooner, reducing system lag to a minimum and providing a narrow differential for the cooling system. Cooling anticipators are provided on all low-voltage cooling thermostats.

### Forced Warm Air System: Nonanticipated Thermostat

If the temperature selector is set on 75 °F (23.9 °C) and the furnace has been off for some time, the temperature in the room drops slowly. The bimetal element follows the air temperature change and closes the electrical contacts at 75 °F. This causes the heating system to start. At this moment no warm air is being delivered to the room because the heating system must warm up to the fan-on setting of the fan control.

While the heating system is warming up, the room air temperature continues to drop slowly. Depending on the type of heating system, the room air temperature will drop to 74.5 ° (23.6 °C) or below before the blower comes on and the warm air is felt by the thermostat bimetal. This difference in temperature between the point at which the thermostat contacts close and the point the air temperature at the thermostat starts to rise in known as *system lag*. The amount of system lag in degrees Fahrenheit depends on the thermostat location and type, the size of the furnace, and the design of the air distribution system.

With the furnace on and the blower running, the room air temperature continues to rise. If the thermostat has a mechanical differential of 2 °F (1.11 °C), the electrical contacts open at 77 °F (25 °C) and shut down the primary heating control on the furnace. However, the furnace is still hot and the blower continues to deliver warm air to the room until the furnace temperature drops to the fan-off setting on the fan control. The additional heat that has been delivered to the room after the thermostat contacts have opened is called *overshoot*. Overshoot can carry the room air temperature to 77.5 °F (25.3 °C) or higher.

### Forced Warm Air System: Anticipated Thermostat

To reduce the wide differential resulting from a nonanticipated thermostat, we simply add a small amount of heat to the bimetal element so that it is slightly warmer than the surrounding room air temperature. We do this by placing a resistor in the thermostat close to the bimetal element. This resistor is in series with the contacts. See Figure 7-20. When the contacts close and the control circuit is energized, the current flowing through the temperature control circuit must also flow through the resistor. The current flowing through the resistor causes it to heat up, which in turn heats the bimetal element. Thus, the point at which the thermostat contacts should open is anticipated to give a narrow differential.

Because the furnace has been on a shorter period of time, there is less heat left in the heat exchanger, which means less overshoot in the space. In the meantime, the bimetal element cools down because it is no longer being heated by the anticipator. Thus, the cycle begins all over again.

System lag and overshoot have not been eliminated, but through the use of anticipation these factors are reduced to the negligible point. Through proper system design and thermostat location, it is not unusual to obtain room temperature differential no greater than $\frac{1}{2}$°F (0.28 °C) by the proper use of heat anticipation.

### Bellows-operated Thermostat

The *bellows-operated thermostat* has the same function as the bimetal thermostat. The major difference between the two is that a bellows filled with some type of fluid that expands when heated and contracts when cooled is used instead of a bimetal element. Bellows-operated thermostats perform the same functions that bimetal thermostats do.

*Operation.*   When the room air temperature begins rising, the fluid temperature inside the bellows also begins rising. As its temperature increases, it begins to expand in direct relation to the temperature. When the cut-in setting of the thermostat is reached, the contacts close and the system starts operating. As the room temperature drops, the temperature of the fluid also drops. When the cut-off temperature is reached, the thermostat contacts open the control circuit and the system stops operating.

### Staging Thermostats

In recent years there has been an increased demand for greater comfort and efficiency in indoor heating and cooling systems. The *staging thermostat* has been designed to meet these needs. See Figure 7-24.

The staging thermostat is designed to be used on systems that have more than one-stage heating and more than one-stage cooling or any combination of heating and cooling stages.

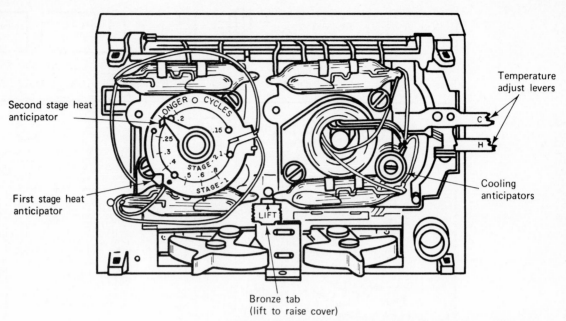

Second stage heat
anticipator

First stage heat
anticipator

Temperature
adjust levers

Cooling
anticipators

Bronze tab
(lift to raise cover)

**Figure 7-24.** Multistage thermostat.

In a typical system using two-staged heating and two-stage cooling, when the thermostat is in the heating position the heating system operates at reduced BTU input capacity during mild weather. As the weather becomes colder, this reduced capacity is not sufficient to maintain the desired comfort level, and the thermostat automatically brings on the additional capacity of the heating system. In the cooling position the situation would be similar, except that we would be bringing on one or two more levels of capacity of cooling rather than heating. These thermostats are also available with automatic changeover. In the automatic position all that is necessary is to set the desired level of heating and cooling, and the thermostat will automatically switch from heating to cooling and back to heating, based on the temperature settings on the thermostat. There are two separate temperature settings on this type of thermostat, one marked $C$ for cooling and the other marked $H$ for heating.

*Circuit Wiring Diagram.*    The circuit wiring with staging thermostats varies from installation to installation due to the number of possible stages that can be used. We shall use a two-stage heating–two-stage cooling wiring diagram to review the thermostat function. See Figure 7-25. From this diagram we can see that the system switch is in the heat position and the fan switch is in the auto position. From one side of the transformer, the circuit goes to terminal RC, through the jumper wire to RH, through the internal wiring to the bar contact on heat, and on to the stage 1 and stage 2 heat anticipators. You will note that there also is a circuit to terminal B, which will energize another circuit continuously. Upon a call for heat, the stage 1 heating switch closes, giving us a circuit to $W_1$, then to the first stage of the heating system, and back to the other side of the transformer. If the temperature continues to drop,

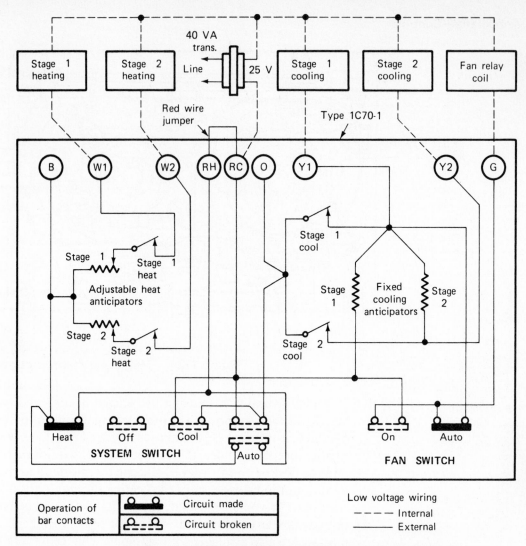

**Figure 7-25.** Multistage system wiring schematic.

the stage 2 heating switch closes, giving a circuit through $W_2$ to the second stage of the heating system and back to the other side of the transformer.

When the selector switch is placed in the cool position, the heat bar contact breaks and the cool contact is made. This gives a circuit from RC to the cool contact, through the internal jumper, to the auto contact, and on to stage 1 and stage 2 contacts. When stage 1 contacts close, there is a circuit through $Y_1$, stage 1 cooling, and back to the other side of the transformer. If this is not sufficient cooling, stage 2 closes and this circuit is made through $Y_2$, stage 2 cooling, and back to the other side of the transformer.

**FAN SWITCHES**  The fan switch on a thermostat has two positions, one for automatic operation and one for continuous operation.

When the fan switch is in the auto position, the fan operates only on demand from that part of the system in use at the time (heating or cooling).

If the switch is placed in the on position, the fan operates continuously regardless of the system demand. The fan also operates in this position when the system switch (*heat* or *cool*) is turned to the off position.

### Modulating Thermostats

Modulating control systems are built around the Wheatstone bridge principle. Both the controller and the controlled apparatus, usually a modulating motor operating a damper or a water valve, use this principle. Both the thermostat (controller) and the motor have potentiometers in them.

The operation of this type of circuit is as follows (see Figure 7-26). If the temperature of the controller rises

1.  The wiper on the controller potentiometer moves toward the W terminal reducing the resistance between W and R at the controller.
2.  Current flow from the transformer through the W terminal of the controller is increased, and this increased current in the corresponding relay coil pulls the DPST relay switch to contact 1.
3.  Current from the transformer then flows through the relay switch 1 on to the motor.
4.  The motor runs counterclockwise to reposition the controlled device. As the motor runs it moves the wiper on the motor potentiometer toward the G terminal.

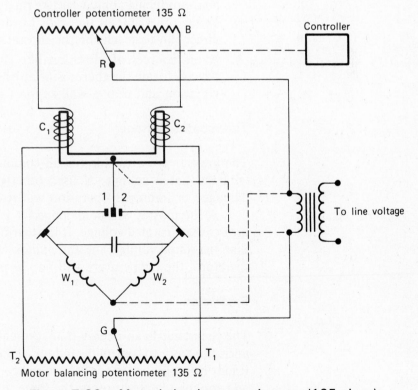

**Figure 7-26.**   Motor balancing potentiometer (135 ohms).

5.  When the wiper on the motor potentiometer reaches a point where the resistance between T and G on the motor potentiometer equals the resistance between R and W on the controller potentiometer, the current relay coil is equalized.

6.  The relay contacts break, stopping the motor. The circuit is again in balance.

On a drop in temperature, the current flows to the other side of the relay and the motor runs in the opposite direction. The potentiometer wiper is moved through a series of levers and springs by a gas-filled bellows.

## Thermostat Location

As was mentioned earlier, the thermostat location is very important for successful operation of the total system.

For your guidance, we offer the following suggestions for proper location of the thermostat:

1.  Always locate the thermostat on an inside wall. If it located on an outside wall, overheating during cold weather is likely to occur because the thermostat will always feel cold.

2.  Avoid the false sources of heat such as lamps, television sets, warm air ducts, or hot water pipes in the wall. Also avoid locations where heat-producing appliances like ranges, ovens, or dryers are located on the other side of the wall. Locations near windows may cause direct sun to reach the thermostat.

3.  Avoid sources of vibration like sliding doors and room doors. Always locate the thermostat at least 4 feet from such sources of vibration and near a wall support if possible.

## Thermostat Voltage

Thermostats are available for all common voltages and must be used with the stated voltage. If used otherwise, they will be permanently damaged or improper operation will result.

A thermostat will be damaged if it is used on a voltage which is greater than its rated voltage. If the thermostat is used on a voltage lower than the stated voltage, the anticipators will not function properly and, therefore, the temperature will not be properly maintained.

## Smartstat

The Smartstat is an energy-management control that makes use of a microprocessor to control the indoor climate and to manage energy usage. It may be programmed for year-round comfort. The user sets the heating and cooling temperatures at the same time and the Smartstat does the temperature-regulation. See Figure 7-27.

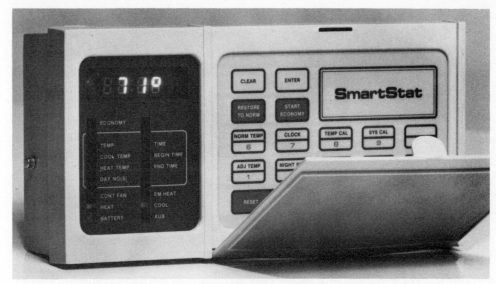

**Figure 7-27.** Smartstat. (Courtesy of Dupont Energy Management)

*Operation.* The Smartstat has three separate programs, which can be used to lower the temperature automatically during the heating season and increase it during the cooling season. This will lower the operating costs when full heating or cooling capacity is not needed. This control computes the amount of time needed for adequate recovery from the setback temperature to the normal setting of the thermostat.

A night/set program is used for economy periods that occur every day of the week on a weekly schedule. A day/set program is used for randomly selected days of the week. A trip/set program is used for economy periods, up to 99 days, that are nonrecurring. The program is equipped with a trip setback mode that lets users set the temperature for times when they are away for extended periods. The memory turns the system back on and brings the temperature to the desired level when they return. The economy periods may be advanced or temporarily cancelled by simply touching the proper place on the front panel of the control.

### Outdoor Thermostat

The outdoor thermostat provides automatic changeover from heating to cooling or cooling to heating in response to the outdoor air temperature. See Figure 7-28. These thermostats either may be bimetal actuated or use a remote bulb for actuating the low-voltage SPDT mercury switch, which breaks one circuit and makes another on a temperature rise or fall in the outdoor air. The operating range of this control is from about 60°F (15.6°C) to about 90°F (32.2°C). These controls are used on systems where accurate temperature control is necessary. The wiring connections for this control are shown in Figure 7-29.

**Figure 7-28**. Outdoor thermostat. (Courtesy of Johnson Controls, Inc., Control Products Division)

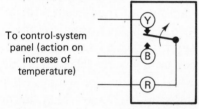

To control-system panel (action on increase of temperature)

**Figure 7-29**. Outdoor thermostat wiring connections.

*Operation.* In operation when the outdoor air temperature rises to the set point of the controller, the mercury bulb will dump, making one circuit and breaking the other. This switches the system into the cooling mode. The switch remains in this position until the outdoor air temperature drops to the set point of the controller; then the mercury bulb dumps in the other direction, stopping the cooling unit and starting the heating unit.

### Fan-Coil Thermostat

*Fan-coil thermostats* are line-voltage controls that are designed to operate heating, cooling, or heating and cooling systems. See Figure 7-30. They are used to control fan motors, relays, or water valves on fan-coil units. They have fan-speed selectors as well as temperature selectors so that the most comfortable conditions can be selected. They have anticipators and bimetal-actuated snap-acting switches included for controlling the equipment. Sequenced models also have a bimetal sensing element with a dead-band (neither heat or cold is provided) incorporated.

**Figure 7-30.** Fan-coil thermostat. (Courtesy of Johnson Controls, Inc., Control Products Division)

*Operation.* Fan-coil thermostats operate the same as any other heating, cooling, or heating and cooling thermostat, except that the equipment cannot be operated for a certain number of degrees on the thermostat between the heating and cooling selector switches.

### Outdoor Heat-Pump Thermostat

*Outdoor heat-pump thermostats* are designed to prevent operation of the auxiliary heating strips during mild weather conditions. See Figure 7-31. These thermostats are SPST NC switches that are mounted in the outdoor unit. The sensing element is located where it will sense the outdoor ambient temperature. The control system to the auxiliary heat strips is interrupted by the opening of the contacts on a rise in temperature. An adjustable temperature range from 0°F (−17.8°C) to 50°F (10.0°C) is provided on most models, which saves energy by not allowing the auxiliary heaters to operate when the heat-pump unit will handle the load.

**Figure 7-31.** Outdoor heat-pump thermostat. (Courtesy of White-Rodgers Division, Emerson Electric Co.)

*Operation.* During mild-weather operation, the outdoor heat-pump thermostat contacts are closed until the outdoor ambient temperature has risen to the cut-out setting of the control and the auxiliary heat strips may be energized as needed. When the cut-out temperature is reached, the contacts open and heat-strip operation is not possible until the outdoor ambient temperature drops to the cut-in setting of the control.

**OUTDOOR RESET CONTROL**

*Outdoor reset controls* are used to maintain a proper balance between heating-medium temperature and outdoor temperature. They automatically raise or lower the temperature of the heating medium (water, steam, or warm air) control point as the outdoor air temperature changes. One sensing bulb is mounted to sense the heating medium and another bulb is used to sense the outdoor air temperature. See Figure 7-32. Outdoor reset controls are equipped with a set of SPDT contacts that change position on a change in temperature. They may be used on line-voltage, low-voltage (24 volts), or millivoltage systems. This control is not designed to replace the safety high-limit control. The differential adjustment of this control is adjustable within the limits of the control.

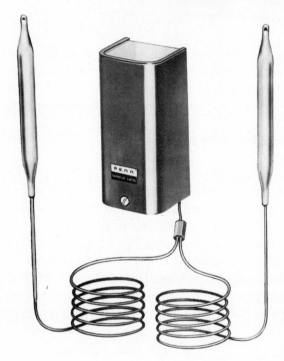

**Figure 7-32.** Outdoor reset control. (Courtesy of Johnson Controls, Inc., Control Products Division)

*Operation.* Because both of the outdoor air temperature and the temperature of the heating medium are measured, a combination of these two temperatures affects the operation of this control. When the outdoor air temperature is high and the system capacity demand is low, the temperature of the heating medium is also low, reducing the amount of energy used to heat the medium. When the outdoor air temperature is low, the temperature of the heating medium is higher to compensate for the additional heat loss from the building.

As the temperature of the outdoor air begins to fall, the outdoor sensing bulb signals the controller that additional heat is needed to maintain the temperature inside the building. The temperature of the heating medium is raised just enough to compensate for the additional heat loss. Likewise, as the outdoor air temperature rises, the outdoor sensing bulb signals the controller that a lower heating-medium temperature will be enough. This lowers the heating cost for the building while maintaining the indoor temperature at the desired level.

**HUMIDISTATS**   *Humidistats* are used to control humidification equipment on air conditioning systems during the heating cycle. It is desirable to add humidity to some structures because the air becomes drier when heating by the heating equipment. Today, there are many different types of humidity-sensitive materials available. For the most part they are organic and include such materials as nylon, wood, human hair, and, in some cases, animal membranes. In addition there are other materials that are sensitive to humidity changes. These materials change in electrical resistance with a change in humidity. Only a few of these are successfully adaptable for use in humidity controllers.

The following is a description of the most commonly used types of controllers and the sensors used in them:

1. *Mechanical-type controls.* Mechanical sensors are dependent upon a change in the length of the size of the sensing element in direct relation to a change in relative humidity. The sensors used in modern controllers are commonly made from human hair or some type of synthetic polymer. They are normally attached to a linkage in the controller, which in turn controls the mechanical, electrical, or pneumatic switching element in a valve or motor.

   In general, they are designed to control at some set point that is adjustable by the user or operator of the equipment. Also available are controllers, which automatically change the set point within the building in response to the outdoor air temperature. This is a step in the reduction of condensation on the windows and building walls.

2. *Electronic-type controllers.* With a change in humidity, the sensors used in electronic humidity controllers also undergo a change in their electrical resistance. These sensors are made of two electrically conductive materials that are separated by a humidity-sensitive hygroscopic insulating material, such as polyvinyl acetate, polyvinyl alcohol, or a solution of certain types of salt.

When air is passed over the sensing element, any small changes in humidity present are detected. The electrical resistance between the two conductors varies in response to the humidity present in the air. In most applications the sensor is placed in one leg of a Wheatstone bridge, where the output signal can be used to control the equipment or as a readout. Generally, a signal amplification is required.

Electronic humidity controllers are most popular in applications where very close control of the humidity is required. In some applications it may be used for controlling the speed of the fan so that a more accurate humidity control can be achieved.

There are several ways to wire these controls into a system. However, they should be wired so that the humidifier operates only when the fan is operating to prevent moisture from collecting on the heat exchanger during the off-cycle. One suggested wiring diagram is shown in Figure 7-33. This diagram provides a safety feature—the water solenoid is not allowed to be actuated unless both sources of electricity are available.

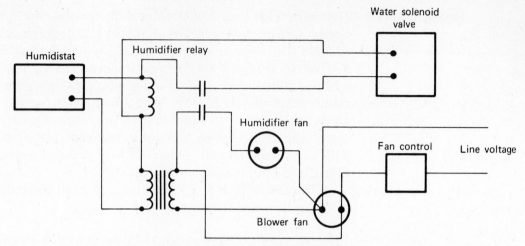

**Figure 7-33.**  Typical humidifier wiring diagram.

Most humidistats may be used on either line-voltage or low-voltage control systems. They usually have a moisture-sensitive nylon ribbon used to actuate the contacts, which open on a rise in humidity. Positive on-off settings are sometimes provided for manual operation. The switch is a SPST, snap-acting type. See Figure 7-34.

**Figure 7-34.** Humidity controller. (Courtesy of White-Rodgers Division, Emerson Electric Co.)

**AIRSTAT**  The airstat controller is used as a fan safety cutoff. It stops operation of the fan whenever the return-air plenum temperature rises to a point that indicates the possibility of a fire. The controller locks out to prevent premature return of fan operation, which would agitate any fire that may be present. The switch may be either bimetal actuated or a fusible link operating a SPST type of control. When this control is actuated, it must either be manually reset or the fusible link must be replaced.

To wire this control into the electrical system, see the diagram in Figure 7-35.

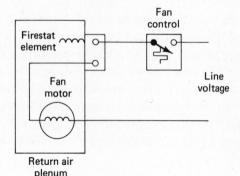

**Figure 7-35.**  Firestat wiring diagram.

**DISCHARGE-AIR AVERAGING THERMOSTAT**

*The discharge-air averaging thermostat* is placed in the discharge-air plenum of an air conditioning system. Its purpose is to cycle the equipment to maintain an average discharge-air temperature. Temperature averaging is a method used to save energy and, therefore, operating expenses. They are generally SPDT switches that can be used to cycle both the heating and cooling equipment to maintain the desired average temperature.

To wire this control into the electrical system, see Figure 7-36.

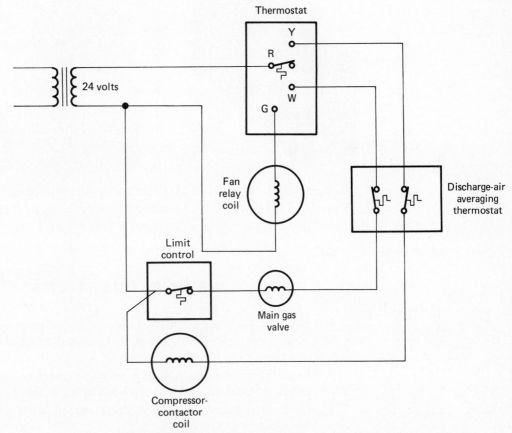

**Figure 7-36.** Typical discharge air-averaging thermostat wiring diagram.

**ENTHALPY CONTROLLER**

The enthalpy controller senses total heat (temperature and relative humidity) of the air. It is designed to control the amount of outside air being brought into the building with respect to the total heat content or enthalpy of the outside air. The humidity sensor generally is a thin nylon ribbon, and the temperature sensor is a liquid-filled bulb that can be mounted in the most desirable position. See Figure 7-37. The enthalpy controller has a SPDT switch with terminals 2 and 1 making on an increase in enthalpy; terminals 2 and 3 make on a decrease in enthalpy below the set point. These controllers can be mounted in any position

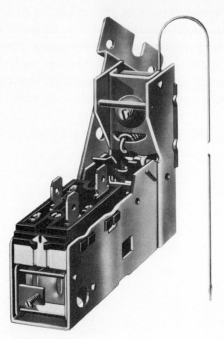

**Figure 7-37.** Enthalpy controller. (Courtesy of Johnson Controls, Inc., Control Products Division)

in an outside air duct. These controllers are equipped with a humidity range of up to 100% and a maximum operating temperature of 150°F (66°C).

## REVIEW QUESTIONS

1. Define temperature controls.
2. What is the major advantage of the space and return-air-sensing method of refrigeration temperature control?
3. Should the thermostat bulb on a space and return-air-sensing control circuit be mounted on a surface which will affect its sensing of the return air?
4. On a refrigeration temperature-control system, when is the alarm circuit energized?
5. What type of refrigeration system control circuit is popular on liquid chillers and soft ice cream machines?
6. What type of temperature control is popular on medium- and low-temperature self-contained refrigeration systems using a capillary tube as the flow-control device?
7. When using the evaporator temperature-sensing device for temperature control, how much of the sensing element must be in direct contact with the surface being sensed?
8. What is the ice-bank control designed to do?
9. With what is the sensing bulb of the ice-bank control filled?
10. Name the two types of thermostat anticipators.
11. What is the purpose of the thermostat anticipator?
12. In what electrical position is the cooling anticipator to the thermostat contacts?

13. In an air conditioning thermostat, heat is added to the bimetal during which cycle?

14. Which system, the anticipated or nonanticipated, provides the most even temperature inside the conditioned space?

15. What needs has the staging thermostat been designed to meet?

16. What is the purpose of the automatic changeover feature on a thermostat?

17. For what is the fan switch on a thermostat used?

18. Around what principle are modulating control systems built?

19. On what wall should the thermostat be installed?

20. What is the purpose of the night/set feature on a room thermostat?

21. In response to what does the outdoor thermostat provide automatic changeover?

22. What are fan coil thermostats used to control?

23. For what is the outdoor thermostat used on heat pump systems?

24. For what are outdoor reset controls used?

25. What type of sensing materials are used in humidistats?

# 8

# Gas Valves
and Pilot
Safety Controls

The automatic control of gas to the main burners of a heating unit is accomplished by the use of electrically powered main-line gas valves. Their safe operation is assured through the use of pilot safety controls in conjunction with a thermocouple, powerpile, or a mercury flame sensor. Automatic operation is accomplished by the use of a thermostat.

**DEFINITIONS**   The following are the accepted definitions of gas valves and pilot safety controls.

**Thermocouple:**   A *thermocouple* is a junction of two bars, wires, or similar devices of dissimilar metals that produce electric current when heated.

**Powerpile:**   A *powerpile* consists of a series of thermocouples connected to produce a higher current than a single thermocouple.

**Mercury Flame Sensor:**   A *mercury flame sensor* is a sensing element that is placed in the pilot flame and filled with mercury to operate a switch in the control circuit.

**Gas Valves:**   *Gas valves* are NC solenoid valves designed for the control of gas to the combustion area of the equipment.

**Pilot Safety Controls:**   *Pilot safety controls* provide safe, automatic shut-off of gas valves during pilot-flame failure in gas-fired appliances.

**THERMOCOUPLE**  A thermocouple is a commonly used source of electric current in which heat is transformed into electric energy. In modern-day heating systems, the thermocouple has become a widely used safety device. It is normally used in conjunction with pilot safety controls. The basic operation of the thermocouple must, therefore, be understood by the service technician.

*Operation.*  When two pieces of dissimilar metals, usually iron and copper, are connected together and heated, an electric current flows. This current is caused by the interaction of the two metals and the resulting movement of electrons. See Figure 8-1.

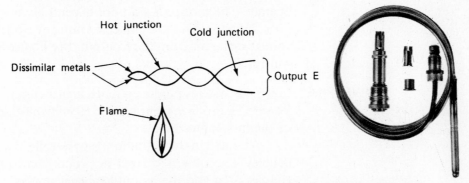

**Figure 8-1.**  (a) Thermocouple schematic, (b) thermocouple. (Courtesy of White-Rodgers Division, Emerson Electric Co.)

The current produced between the hot and cold junctions increases in temperature up to a given current rating. Above this level there will be little, if any, rise in current regardless of the amount of heat applied. These devices should not be overheated, because excessive temperatures will result in a burned-out junction and a ruined thermocouple.

Thermocouples are normally used as safety devices and will hold an electromagnetic relay in the pulled-in position after the armature is manually pushed into the electromagnetic coil. If, for some reason, the current is reduced or stopped, the armature will be pulled out of the coil by spring pressure.

Thermocouples are rated at 30 millivolts when the end is inserted into the pilot flame from $\frac{3}{8}$ to $\frac{1}{2}$ inch. See Figure 8-2.

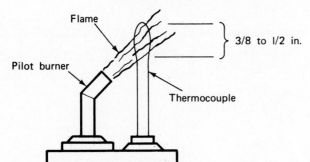

**Figure 8-2.**  Relationship of pilot flame and thermocouple.

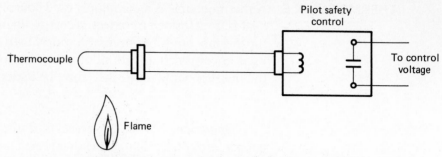

**Figure 8-3.**  Thermocouple regulated control.

These devices are connected directly to the individual controls they operate. Thermocouples do not depend on any outside voltage to perform and should not be connected into any electrical-control circuit with voltage different from their output. See Figure 8-3. A damaged control will result.

**POWERPILE**  Powerpile systems produce a much higher current than do thermocouples. Powerpiles produce from 150 to 750 millivolts, depending on the number of thermocouples used.

As stated in the definition, a powerpile is simply a series of thermocouples. It produces current in proportion to the number of thermocouples used and the amount of heat applied. See Figure 8-4.

Powerpile systems are usually used to produce the full operating current for a millivolt control circuit. See Figure 8-5. As with a single thermocouple, this kind of system must be separated from any other electric current.

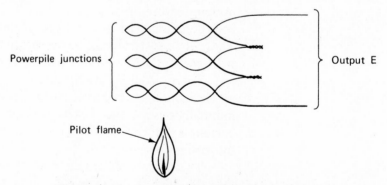

**Figure 8-4.**  Fundamental powerpile element.

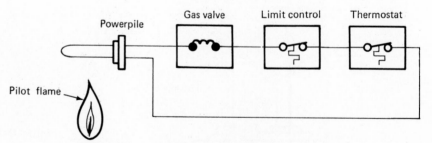

**Figure 8-5.**  Typical millivoltage wiring diagram.

For proper operation of the heating equipment, the proper thermostat or one with an adjustable heat anticipator must be used. All electrical connections should be scraped and soldered to reduce voltage drop to a minimum.

*Caution:* Do not use a 30-millivolt control with powerpile voltage because a burnout of the control will result, similar to the result when using 120-volt controls on a 230-volt source.

**MERCURY FLAME SENSOR**    Mercury flame sensors consist of a bulb, a capillary tube, and a diaphragm assembly, which are all connected to a snap-acting switch. The mercury bulb is inserted into the pilot burner flame to sense if the flame is lit or not. When the pilot flame is proven, the snap switch energizes the main gas-valve relay. See Figure 8-6. These sensors are also available for control circuit interruption only. See Figure 8-7.

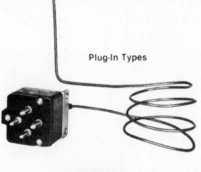

Plug-In Types

**Figure 8-6.**   Plug-in type mercury flame detector. (Courtesy of White-Rodgers Division, Emerson Electric Co.)

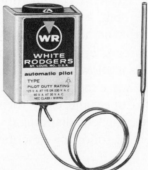

**Figure 8-7.**   Switch-type mercury flame-type detector. (Courtesy of White-Rodgers division, Emerson Electric Co.)

*Operation.*   When the pilot is proven the mercury bulb is heated, causing the mercury to expand. When a predetermined amount of pressure is exerted on the diaphragm, a snap-acting switch is closed, completing the electrical circuit to the main gas valve. If the pilot goes out or becomes too small to ignite the main burners adequately, the pressure inside the diaphragm decreases because of the lower temperature. This allows the snap switch to open the control circuit to the main gas valve and shut down the main burners.

**GAS VALVES**    The main gas valve is a device that acts in accordance to the demand of the thermostat either to open or to close. This valve is designed to provide many functions. Its components are a gas-pressure regulator,

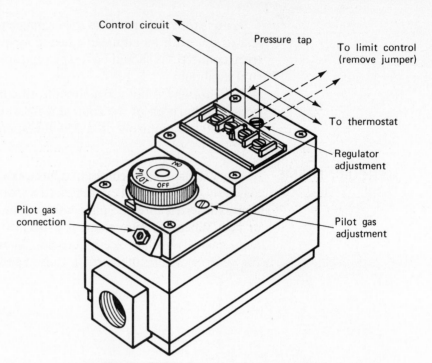

**Figure 8-8.** Combination gas valve.

a pilot safety, a main gas cock, a pilot gas cock, and the main gas valve. These components are all contained in a single valve body that is generally referred to as the *combination gas control.* See Figure 8-8. The purpose of the main gas valve is to admit gas to the main burners on demand from the thermostat. These valves are usually of the globe type.

The main parts are the valve body, valve seat, valve disc, and the valve stem. Gas is allowed to flow to the equipment when the valve disc is lifted off the valve seat. Gas valves are classified by the type of power that is used to open and close them. They are classified as follows: solenoid valve, magnetic diaphragm valve, bimetal diaphragm, and heat motor valve. The following describes each of these valves.

### Solenoid Valve

The *solenoid valve* is generally considered to be an electrically operated stop valve. It is normally closed and is the most popular type used on heating systems.

*Operation.* When the electrical circuit is completed to the solenoid coil, the valve opens. When the circuit is interrupted, the valve closes.

The noise made by the closing of the solenoid valve is lessened by the addition of a rubber or plastic seat in the valve. This, in turn, is given the name *soft-seat solenoid.* A spring is added above the valve to slow down the opening and to help stop the clacking noise made when the valve opens.

An improvement on this valve is the oil-filled solenoid valve. The oil reduces friction and the operating noise is almost completely eliminated.

### Magnetic Diaphragm Valve

The *magnetic diaphragm valve* makes use of the available gas pressure as the major opening and closing force.

*Operation.* On demand from the thermostat, the control circuit energizes the valve coil. When the valve is energized a lever is magnetically pulled, closing a gas supply passage. At the same time the gas pressure above the diaphragm is vented out through the bleed orifice and to the pilot burner, where it is immediately burned. The gas pressure below the diaphragm is now greater than that above the diaphragm, and the weight and disc are lifted off their seat, opening the gas valve.

When the control circuit is de-energized, the bleed gas is stopped, allowing the gas pressure above and below the diaphragm to equalize. The weight of the valve disc closes the valve.

### Bimetal Diaphragm Valve

The *bimetal diaphragm valve* and the magnetic diaphragm valve are identical in operation except that a bimetal is used to open and close the bleed orifice.

*Operation.* The bimetal diaphragm valve uses the principle of dissimilar metals, which warp when heated to open and close the bleed orifice. An electric heater is placed either on or near the bimetal element. When the control circuit is completed to the heating coil and sufficient heat is added to the bimetal, the bimetal will warp. The bimetal opens the bleed orifice and the gas pressure is allowed to vent to the pilot burner. This reduces the pressure above the diaphragm and allows the greater pressure below the diaphragm to force the valve open.

When the control circuit is de-energized, the bimetal cools, returns to its original shape, and closes the bleed orifice. This allows gas pressure to build up above the diaphragm and the seat. The weight of the diaphragm and seat closes the valve.

### Heat Motor Valve

The *heat motor valve* uses the heat given off by an electrical heating coil to move an expandable rod to open the valve. The movement of this heated rod provides the force necessary to operate the valve mechanism.

*Operation.* When the thermostat demands heat, the control circuit to the heating coil is energized and the coil is heated. The heat given off by the coil causes the expandable rod to expand and the valve to snap open.

When the thermostat is satisfied, the control circuit is opened and the coil is allowed to cool. When the rod has cooled sufficiently, it shrinks in length and the valve snaps closed.

*Note:* Most of these valves have a certain amount of time delay before they open or close. This is to provide a smoother ignition and extinguishing of the flame.

### Redundant Gas Valve

The purpose of *redundant gas valves* is safety. There are two independently operated valves that must open before gas is admitted to the main burner. If one of these valves should fail to close, the other one closes and stops the flow of gas to the main burner. See Figure 8-9. This type of valve is used on gas-fired heaters and boilers, with or without intermittent pilot ignition, in place of the regular gas valve.

*Operation.* In the majority of cases, any problems with this type of valve are electrical rather than mechanical. They have two shutoff valves that are electrically operated and in series with each other. The

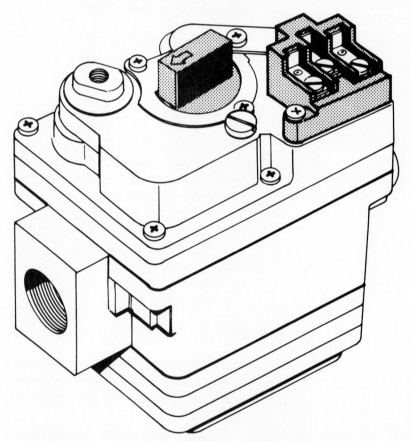

**Figure 8-9.** Redundant gas valve. (A Reston Publication, reprinted by permission of Prentice-Hall, Englewood Cliffs, N.J.)

flow of gas does not occur until both of these valves are open. Only one needs to close to stop the flow of gas, however. This is a safety feature, because it is almost impossible for both valves to stick in the open position at the same time.

There are several different types of redundant gas valves. One type uses an instant-acting solenoid valve at the valve outlet to control the flow of gas. It also uses a time-delay valve at the inlet of the valve. This time delay is usually about 10 seconds, providing a time delay on the opening of the gas valve.

The type of redundant gas valve used on gas furnaces with standing pilots is similar to the standard type of gas valve except that an internal heat motor valve is also used. Thus, there are two 24-volt coils in these valves. See Figure 8-10.

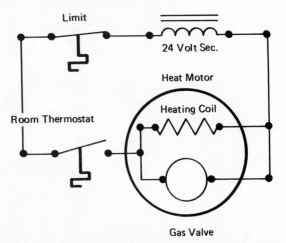

**Figure 8-10.** Two coils in one valve. (A Reston Publication, reprinted by permission of Prentice-Hall, Englewood Cliffs, N.J.).

In this diagram one coil is shown as a solenoid valve and the other coil is shown as a resistor. They are in parallel electrically. Both of them are in series with the limit control and the thermostat. The resistor indicates the heat motor that operates the second, or redundant, gas valve.

There is a third internal valve, the 100% safety shutoff valve. It is energized by the electricity produced by the thermocouple (millivolts). If the pilot flame is not sufficient to heat the thermocouple, the 100% safety shutoff will close stopping the flow of gas to the main burner.

If either the limit control or the thermostat opens the control circuit, the solenoid valve closes instantly. However, the heat motor valve takes a few seconds to cool down and close the second valve.

## Automatic Spark-Ignition Gas Valves

*Automatic spark-ignition gas valves* are equipped with a redundant pilot solenoid, main gas regulator, and the necessary wiring connections for the make and model valve being used. They use electrically operated

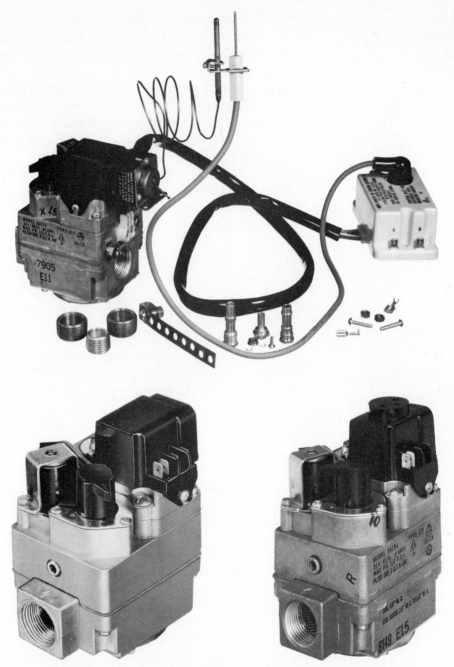

**Figure 8-11.** Automatic spark-ignition system main gas valves (Courtesy of White-Rodgers Division, Emerson Electric Co.).

solenoid valves controlled by a room thermostat or other temperature control. See Figure 8-11.

*Operation.*    The main gas valve is controlled by the proper flame sensor, which plugs into the valve. Both the pilot and the main gas valve are controlled through the proper relight-timer control, which de-energizes both valve coils if a pilot flame is not established within the lockout period. These valves should be installed only on systems that are equipped with the proper ignition systems as recommended by the manufacturer for proper operation.

**PILOT-SAFETY CONTROLS**     Pilot-safety controls are subdivided into two distinct types: (1) thermopilot valves and (2) thermopilot relays.

The thermopilot valve is a 100% safety shut-off control used for gas-fired appliances and heating equipment. A push button provides the manual reset operation. See Figure 8-12. This valve is installed in the main gas line ahead of the main gas valve. See Figure 8-13.

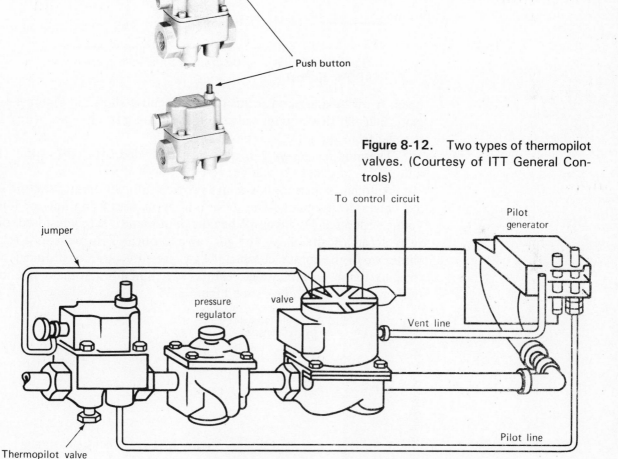

**Figure 8-12.** Two types of thermopilot valves. (Courtesy of ITT General Controls)

**Figure 8-13.** Installation of thermopilot valve.

*Operation.* Electric current from the thermocouple energizes an electromagnet to hold the valve open after it is manually reset. Loss of current to the electromagnet due to low pilot flame, pilot-flame outage, or limit control operation causes the electromagnetic valve to snap closed, shutting off the gas flow. Depressing and holding the button opens the gas valve to allow pilot ignition. After the pilot burns about 60 seconds, electrical current from the thermocouple is sufficient to hold the valve

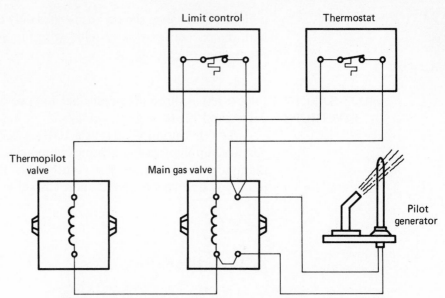

**Figure 8-14.**  Wiring diagram for millivolt system with 100% shutoff.

open. A wiring diagram for this type of circuit is shown in Figure 8-14. Note that the thermopilot valve and the main gas valve are wired in parallel from the pilot generator.

Thermopilot relays can also be subdivided into two types: (1) automatic reset and (2) manual reset.

Automatic thermopilot relays provide safe, automatic shutoff of gas valves during pilot-flame failure in appliances using line- or low-voltage-operated gas valves. When the pilot flame is safely restored, the relay automatically allows the gas valve to open. This prevents a false shutdown of the appliance when the gas pressure varies or drafts affect the pilot flame. The relay also features switching for automatic ignition during pilot-flame failure. See Figure 8-15.

Thermocouple connection

**Figure 8-15.**  Thermopilot relay. (Courtesy of ITT General Controls)

The relay is powered by millivoltage from the pilot generator, which is heated by the pilot flame. This energizes the relay and closes its contacts to complete the electrical circuit to the main gas valve, allowing gas to flow. See Figure 8-16.

If insufficient millivoltage is produced by the heat of the pilot flame, the relay is de-energized; the relay opens its contacts and the circuit to the main gas valve is broken. The main gas valve then shuts off the gas.

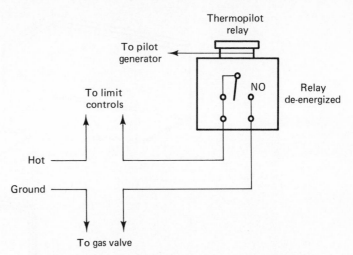

**Figure 8-16.** Thermopilot relay wiring diagram.

In SPDT models, the de-energized relay also completes a second circuit, which can be used for ignition or signal alarm during pilot-flame failure. See Figure 8-17.

The manual-rest thermopilot relay is an electric contacting device that provides safe lighting, automatic safety shutoff for gas appliance controls, and a manual reset indicator.

A reset on all manual-reset relays shows if the relay is on or off; this simplifies lighting operations and permits manual shutoff for the appliance. A pilot gas valve in a single unit provides safety shutoff control of the pilot gas in addition to main gas shutoff if the pilot is extinguished.

Electrical power for operating the electromagnetic relay is supplied by a single thermocouple (30 millivolts). On SPST models, turning the knob to pilot reset opens the relay contacts. If a pilot valve is also incorporated, gas also flows to the pilot. After the pilot has been burning for about 1 minute, the reset can be released and allowed to turn to 'on'. See Figure 8-18.

If the pilot flame becomes unsafe, the relay contacts automatically open, the pilot valve closes, and the reset turns to 'off'. The system can be shut off at any time by turning the reset to off.

**Figure 8-17.** De-energized SPDT relay schematic.

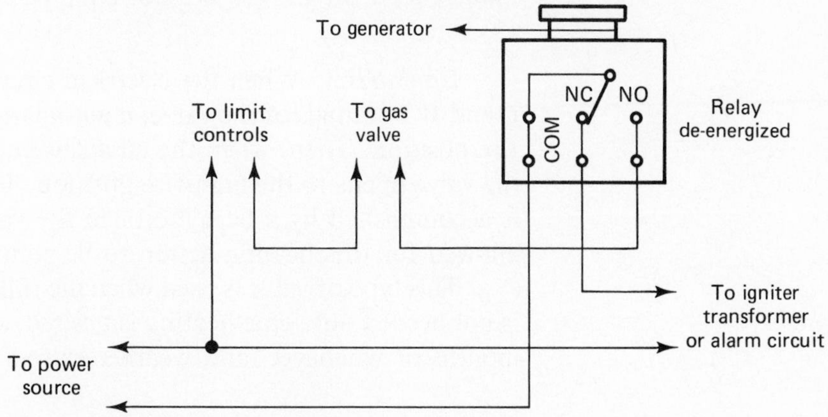

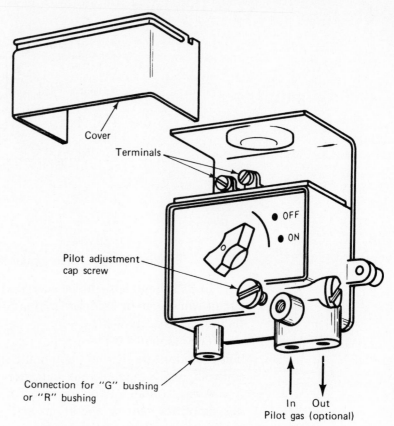

**Figure 8-18.** Manual reset thermopilot relay.

**HIGH-LOW FIRE GAS VALVES**

The *high-low-off main gas valves* provide all manual and automatic control functions required for operation of gas-fired heating equipment. They have an internally vented diaphragm-type main control valve and a separate thermomagnetic safety valve with pilot gas adjustment and a pilot filter. Most models include a pressure-regulator function for use with natural gas. See Figure 8-19.

The main gas-valve diaphragm opens and closes in response to a thermostat and/or a limit control. The main-line diaphragm valve functions simultaneously as a pressure regulator on natural gas models. The models used on LP gas are not equipped with a regulator.

*Operation.* When the electrical circuit is completed between the $C$ and $W$ terminals of the valve, it automatically opens to the preset low-fire position. Then, when the circuit is completed between $C$ and $W_2$, the valve opens to the high-fire position. The low-fire to high-fire shift is accomplished by a heat motor in the valve operator. Time must be allowed for this heating action to be completed. See Figure 8-20.

This type of valve is used when the full BTU rating of the appliance is not needed but some heating is desired, such as in the spring and fall months or whenever mild weather exists.

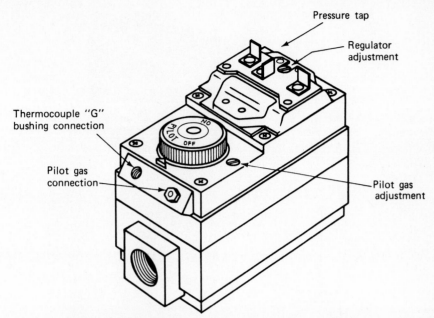

**Figure 8-19.** High-low-off combination gas valve.

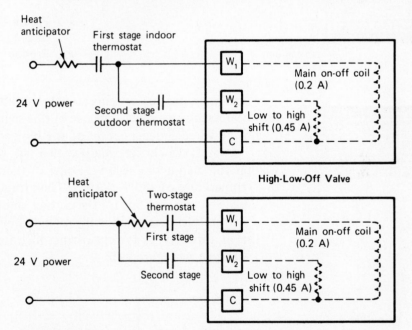

**Figure 8-20.** Schematic diagram for high-low-off combination gas valve.

**SAIL SWITCH**    *Sail switches* are designed to detect air flow or the lack of air flow in ducts. They respond only to the velocity of air movement. Their purpose is to activate electronic air cleaners, humidifiers, gas valves, or other auxiliary equipment in response to airflow from the system's circulating air fan. They are also used as safety devices on forced-draft gas burners to prevent opening of the main gas valve until sufficient combustion air is supplied to the combustion chamber. Some manufacturers use them

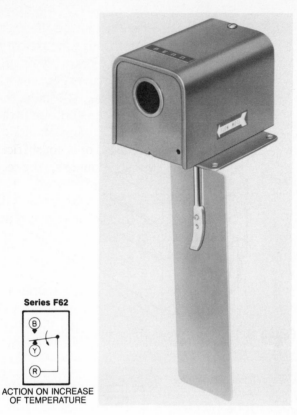

**Series F62**

ACTION ON INCREASE
OF TEMPERATURE

**Figure 8-21.** Sail switch (Courtesy of Johnson Controls, Inc., Control Products Division).

on electric heating systems to prove a minimum air flow before the strip heaters are energized. The use of this switch allows auxiliary equipment to be wired independently of the blower motor. See Figure 8-21.

Such switches are made with a sail, which is mounted on an SPDT, micro, snap-acting switch. Some of them have a maximum operating temperature of 180°F (82°C). Therefore, when replacing a sail switch, make certain that one for the correct ambient temperature is being used. Some of them are positionally mounted; use caution to install the replacement correctly. A typical wiring diagram for an electronic air cleaner is shown in Figure 8-22.

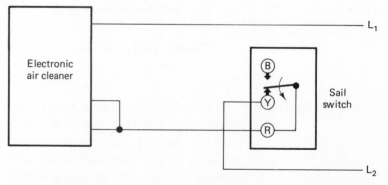

**Figure 8-22.** Typical sail switch schematic for an electronic air cleaner.

*Operation.*    The thermostat generally controls the operation of the fan motor, which in turn moves the air through the system as required. The operating contacts of a sail switch are normally open. As the fan starts blowing air through the area being measured, the sail is caused to move. When sufficient air is flowing through the area, the sail moves far enough to close the contacts in the switch. At this time the controlled equipment is energized, and the system functions as designed.

If for any reason the air flow is reduced or is not sufficient to maintain the closed condition of the NO contacts, the equipment is automatically shut down.

When the thermostat is fully satisfied, the fan motor is de-energized, stopping the air flow through the system. As the flow is reduced the sail switch moves to open the operating contacts which, in turn, de-energize all the equipment under its command.

**AUTOMATIC FLUE DAMPER**    Any time that a gas furnace is in operation, air is drawn from inside the building and is vented to the atmosphere, along with the products of combustion, through the vent system. When the burners are not lighted, this air flow is not needed; however, air continues to flow as long as there is a temperature difference between the inside and the outside of the building. This air carries the heated air out with it, representing a waste of energy. The automatic flue damper is one means of reducing this waste of heated air. This reduction is accomplished by opening and closing the vent damper in response to the thermostat.

A savings of as much as 16% can be realized by installing the automatic flue damper on the top of the furnace-draft diverter and then connecting the venting system to the damper. See Figure 8-23.

**Figure 8-23.**   Location of automatic flue damper (Courtesy of Carrier Corporation).

*Operation.*   On a demand for heat from the thermostat, the automatic flue damper moves to the full open position before the burner can be ignited. See Figure 8-24. The damper blade remains in the full

**Figure 8-24.**  Automatic flue damper open (Courtesy of Carrier Corporation).

**Figure 8-25.**  Automatic flue damper closed (Courtesy of Carrier Corporation).

open position while the burner is on. In the advent of a power failure, a spring-loaded mechanism moves the damper to the full open position for safety.

When the thermostat is satisfied, an electric motor automatically closes the automatic flue damper to stop the flow of air up through the venting system. See Figure 8-25.

# REVIEW QUESTIONS

1. In what device is heat transformed into electric energy?
2. For what purpose are thermocouples normally used?
3. What, other than temperature, determines the output of a powerpile?
4. What device is used to produce the full operating current for a millivolt control circuit?
5. What produces the operating force in a mercury flame sensor?
6. What is the action of the main gas valve?
7. What is the most popular type of gas valve used on heating systems?
8. What is the major opening and closing force used in magnetic diaphragm gas valves?
9. What is the purpose of the redundant gas valve?
10. In what type of gas valve is trouble likely to be electrical rather than mechanical?
11. What type of gas valve is controlled by the proper sensor plugged into the valve?
12. Name the two types of pilot safety controls.
13. For what purposes are sail switches designed?
14. What is the purpose of the automatic flue damper?
15. Where is the automatic flue damper installed?

# 9

# Fan Controls and Limit Controls

In the days of hand-fired coal furnaces and boilers, the person who did the shoveling waited around long enough to make sure that the fire was burning properly. That is just one of the many jobs performed today by automatic controls. These silent servants are on duty around the clock to operate heating and cooling systems safely and economically and to provide comfortable living conditions. Two more such servants are the fan control and the limit control.

**DEFINITIONS**

The following are the accepted definitions of the fan and limit control.

**Fan Control:**  The *fan control* is a device used to start and stop the fan in response to the temperature inside the furnace-heat exchanger. A fan control has NO SPST contacts, which are actuated by temperature. The control is mounted to sense the air temperature within the circulating air passages of the furnace heat exchanger.

**Limit Control:**  The *limit control* is a safety device used to interrupt the control circuit to the main gas valve in case of excessive temperatures inside the furnace. The limit control has NC contacts. It may be either an SPST or an SPDT switch actuated by temperature. It is mounted to sense the air temperature within the furnace.

**FAN CONTROLS**

Forced-air circulation systems have controls known as *heat watchers* in addition to the burning controls. These additional controls permit the circulating fan to operate only when there is enough heat available to heat the conditioned space.

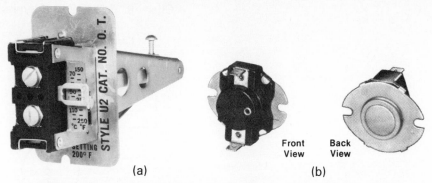

(a)                                    (b)

**Figure 9-1.** Heat-actuated fan controls. (a) Blade type and (b) disc type. (Courtesy of White-Rodgers Division, Emerson Electric Co.)

**Figure 9-2.** Electrically operated fan control. (Courtesy of White-Rodgers Division, Emerson Electric Co.)

There are two types of fan controls: (1) temperature-actuated (see Figure 9-1) and (2) electrically operated (see Figure 9-2).

### Temperature-actuated Fan Control

The *temperature-actuated fan control* uses either a bimetal strip, a bimetal disc, a helical bimetal, or a pneumatically operated sensing element. The sensing element is inserted into the warm-air plenum or directly into the circulating air passage of the furnace heat exchanger. In either case these elements are mounted in such a position that they sense the temperature of the discharge air from the furnace. See Figure 9-1. They are available with sensing elements of different lengths so that the temperature can be sensed in the exact place that the manufacturer has determined to be the hot spot of the unit. They are adjusted by a sliding lever on a scale that is either printed or etched on the control element dial.

*Operation.* In operation, when the thermostat calls for heat, the main burner gas is ignited, and the furnace heat exchanger begins to warm up. As the temperature of the furnace increases, the sensing element begins to respond and move the dial toward the "on" setting of the control. When the "on" setting of the control is reached, usually between 125 °F (51.5 °C) and 150 °F (66 °C), the snap action of the switch closes the NO contacts and completes the line voltage circuit to the fan motor. See Figure 9-3.

When either the thermostat is satisfied or the limit control contacts open, the main gas valve closes, stopping the flow of gas to the main burners. As the furnace cools down, the air being delivered to the conditioned space also cools down and causes the sensing element to return

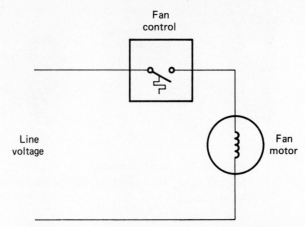

**Figure 9-3.**   Schematic of fan motor wiring.

to its original position. When the sensing element has returned to its original position, the switch contacts are opened and the fan motor stops. The "off" temperature of the fan is usually about 100 °F (37.8 °C).

### Electrically Operated Fan Control

The *electrically operated fan control* (fan timer control, see Figure 9-2) provides a timed fan operation for forced-warm-air furnaces when the heater coil is wired in parallel to the low-voltage coil of the main gas valve. See Figure 9-4. This control is particularly suited for counterflow and horizontal furnaces because of the peculiarity of heat buildup in the furnace after the main gas valve is closed. The heater coil uses 24 volts ac. The switch is an SPST, heater-actuated, bimetal control. The fan is usually started about 1 minute after the thermostat calls for heat; the fan is stopped about 2 minutes after the thermostat is satisfied. This type of control causes the fan to operate even when there is no flame in the furnace because it is electrically actuated rather than flame actuated.

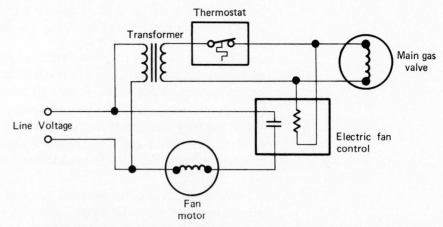

**Figure 9-4.**   Schematic diagram of an electrically operated fan control.

*Operation.* When the thermostat demands heat, the heater inside the fan control and the gas valve are energized at the same time. After the predetermined period of time has lapsed, the fan-control contacts close and start the fan motor. Theoretically, the main gas valve has also opened and the furnace has warmed up sufficiently to warm the building by this time. When the thermostat is satisfied, the main gas valve and the fan control are de-energized at the same time. The flame is extinguished immediately, and the fan continues to run and cool the furnace down. After about 2 minutes the fan motor is de-energized and the furnace is ready for the next operating cycle.

## LIMIT CONTROLS

**Figure 9-5.** Limit control. (Courtesy of White-Rodgers Division, Emerson Electric Co.)

The limit control is used on all types of warm-air furnaces to prevent excessive plenum temperatures and a possible resulting fire. They have either NC SPST switch contacts or SPDT switch contacts, depending upon their specific uses. The contacts on the SPST type open on a rise in temperature and close on a fall in temperature. The SPDT contact types open the NC contacts on a rise in temperature and close the NO set at the same time. The NO set is used to energize the fan motor on horizontal and counterflow type furnaces when excessive temperatures exist inside the furnace. The snap-action switch may be actuated by either a bimetal strip, a bimetal disc, a helical bimetal, or a pneumatically operated sensing element. The sensing element is placed so that it will sense the air temperature inside the furnace heat exchanger. The sensing elements are available in different lengths. Be sure to use the correct length so that proper operation of the equipment will be maintained. See Figure 9-5.

Some models are adjustable and others are nonadjustable. Those that are adjustable have a range from about 180 °F (82 °C) to 250 °F (121 °C) with a fixed differential of 25 °F (13.9 °C). They are adjusted by moving a sliding lever along a scale that is printed or etched on the control element dial. They usually have a stop to prevent setting the "off" temperature higher than 200 °F (93 °C). The "on" temperature is 25 °F lower than the off temperature setting. Only the cut-out is adjusted.

*Operation.* In operation, when the off setting is reached, the electrical circuit to the main gas valve is interrupted. This can be done either by opening the 24-volt ac control circuit directly (Figure 9-6) or by interrupting the line voltage to the primary side of the transformer. See Figure 9-7. This interruption allows the main gas valve to close, stopping the flow of gas to the main burners. As the furnace cools down the sensing element also cools. When the predetermined on setting is reached, the electrical circuit is completed to the main gas valve and operation is resumed.

When the SPDT type is used the operation is a little different. When the NC contacts open, the NO contacts close. At this time the fan motor is energized to blow the hot air from the furnace. When the furnace has cooled down, the contacts switch, stopping the fan motor and energizing the main gas valve.

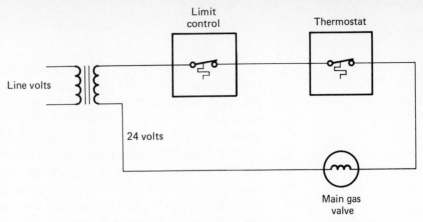

**Figure 9-6.** Limit control in 24-volt ac control circuit.

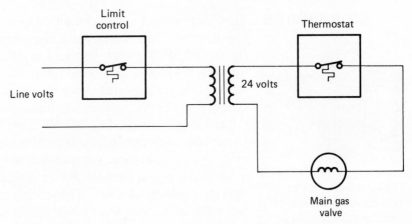

**Figure 9-7.** Limit control in line voltage to primary side of transformer.

This control is designed to function only when there is an excessive amount of heat inside the furnace. The SPST types are usually designed for pilot duty only and will not accomodate heavy current draw through the contacts. However, the SPDT type has contacts rated heavy enough to carry the fan motor current.

**COMBINATION FAN AND LIMIT CONTROL**

Combination fan and limit controls provide fan and limit control on forced-warm-air furnaces and heaters. They are manufactured with a snap-acting, sealed switch. The contacts are designed for millivolt, 24-volt, 120-volt, or 240-volt circuits.

*Operation.* The limit control interrupts the control circuit on a rise in temperature to stop operation of the main burner if the plenum temperature reaches the off setting. This is the same action as that provided by the single-limit control discussed earlier.

On a temperature rise, the fan switch completes the electric circuit to start the fan motor when the plenum temperature rises to the on setting. The fan motor stops when the plenum temperature drops to the fan off setting. This is identical to the operation of the single fan control discussed earlier.

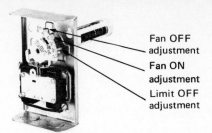

Fan OFF
adjustment

Fan ON
adjustment

Limit OFF
adjustment

**Figure 9-8.** Combination fan and limit control. (Courtesy of ITT General Controls)

The combination fan and limit controls combine the functions of the individual fan and limit controls into a single compact unit. The scale setting is simplified. The limit action cannot be set below the fan-control setting. On some models a summer fan switch is readily accessible without removing the cover and provides for the selection of continuous fan operation or automatic fan operation. See Figure 9-8.

The two switches in these controls are never wired in series. Always wire them in parallel. The limit control may be either in the low-voltage circuit or it may be in the line-voltage circuit, depending upon the desires of the equipment manufacturer. The fan control is always wired into the line voltage circuit so that it can control operation of the fan directly. For some suggested wiring diagrams, see Figures 9-9 and 9-10.

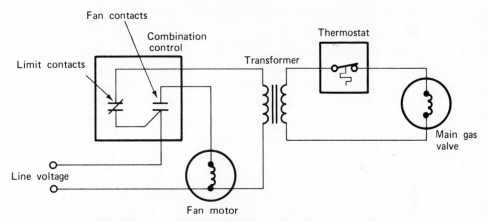

**Figure 9-9.** Typical control circuit using combination fan and limit control.

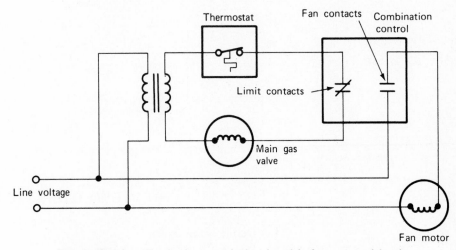

**Figure 9-10.** Typical control circuit with fan control in the line-voltage circuit and limit control in low-voltage circuit.

**HIGH-LIMIT
CONTROLLER**
High-limit controllers are safety switches that are used to interrupt either the line voltage or the low-voltage circuit and stop the fan if the air temperature reaches a preset temperature. They provide positive lockout of the main burners in case of fan-motor failure. They are manually reset to prevent operation of the equipment when it is in an unsafe condition. A bimetal strip inserted directly into the air stream actuates an SPST, NC switch. See Figure 9-11.

5C06-125

**Figure 9-11.** High-limit controller. (Courtesy of White-Rodgers Division, Emerson Electric Co.)

**AIR SWITCH**
*Air switches* are used to control the fan in warm-air applications to prevent reverse air circulation and excessive filter temperatures in counterflow furnaces. Two-speed fan operation is obtained through an SPDT switch. The bimetal is inserted directly into the air stream. See Figure 9-12. These controls are designed for pilot duty only with a maximum current flow of 50 volt-amps at 24 volts ac.

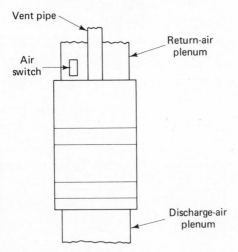

**Figure 9-12.** Air switch location.

## REVIEW QUESTIONS

1. Define a fan control.
2. Define a limit control.
3. On a fan control, where is the sensing element positioned?

4. What is generally the on setting temperature of the fan control?

5. How is the heater coil in an electric fan control wired into the circuit in relation to the main gas valve?

6. To what type of furnaces are electric fan controls particularly suited?

7. Do the contacts on a limit control open or close on a rise in temperature?

8. Generally, what is the highest possible temperature setting on a limit control?

9. In combination fan and limit controls, how should the two switches be wired in relation to each other?

10. What is the purpose of the high-limit controller?

# 10

# Automatic Gas-Burner Ignition

The gas burner is the device used for mixing the gas and air for proper burning. There are two types used in modern gas heating furnaces: (1) main burner and (2) pilot burner. The pilot burner is used for igniting the main gas burners and to provide the heat required to operate the pilot-safety devices in gas-heating furnaces. The main burners generally operate in on and off modes, requiring relighting each time the thermostat demands heat. Until recently gas burners primarily used standing pilots; however, more and more equipment manufacturers are beginning to make use of intermittent pilot-ignition methods. Standing pilots may be lighted manually or by an automatic pilot-ignition control, but intermittent pilots must be lit each time the thermostat demands heat. A standing pilot is one that burns all the time after being lit, and an intermittent pilot is one that burns only when there is a demand for heat. Intermittent pilots are used to conserve energy and for convenience to the user. There are several methods of relighting the pilot, such as spark ignition, glow coil, and other methods limited only to the imagination.

**DEFINITION**   The automatic gas-burner ignition system automatically and instantly lights or relights the gas burner as required for that particular installation.

**TYPES**   There are two general types of automatic gas-burner ignition systems: the glow coil and the spark, or electronic, ignition systems.

136

### Glow Coil

The *glow coil* type is generally used for lighting a standing pilot. See Figure 10-1. This unit incorporates a purge-time delay of approximately 5 minutes if the pilot flame should go out. This delay allows any gas that has accumulated in the furnace to escape. Relighting the pilot, when required, is automatic at any time that 24-volt ac power is supplied to the unit.

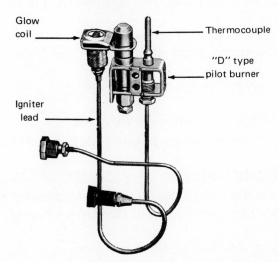

**Figure 10–1.** Glow coil and pilot. (Courtesy of Control Products Division, Johnson Controls, Inc.)

*Operation.* The following is a description of the operating sequence of a glow coil system.

*Normal Lighting Sequence:* Refer to Figure 10-2.

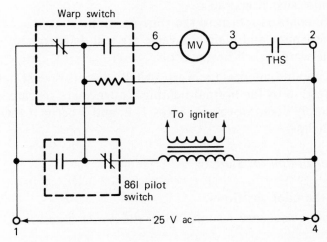

**Figure 10-2.** Schematic wiring diagram.

1. A and B valves are open (not shown). Only the electrical circuit is shown in this diagram.
    a. Pilot gas flows to the pilot burner.
2. Power is available to the control.
    a. The warp switch is heated.
    b. The ignitor is energized.

3. Pilot-burner gas ignites.
   a. The thermocouple is heated.
4. Pilot switch operates (usually about a 20-second delay).
   a. The NC contacts open; the NO contacts close.
   b. The ignitor is de-energized.
5. The warp switch operates (approximately 2 minutes after Step 4).
   a. The thermostat and main-valve circuit now have electrical power.
6. The thermostat contacts close on a call for heat; the main gas valve opens.
   a. The main burner gas ignites and the system now operates.

*Note:*   Step 6 cannot occur unless steps 4 and 5 are completed.

*Safety Sequences:*   No gas on start.

1. Warp switch and ignitor are energized.
2. The warp switch operates (about 2 minutes later) to de-energize the ignitor.
   a. The warp switch now cools.
3. When the warp switch cools (about 5 minutes) another ignition try is made. This continues until the electrical circuit is interrupted.

*Interruption of Gas During Running Cycle:*

1. The pilot and main burner flames are extinguished.
   a. The gas supply may return without ignition.
2. The thermocouple cools in about 45 seconds.
3. The pilot switch operates.
   a. The contacts return to the start position.
   b. The ignitor and main gas valve are still de-energized.
   c. The warp switch heater cools.
4. After about 5 minutes the warp switch contacts reverse and Steps 2 through 6 of the normal lighting sequence are repeated. If the gas supply does not return, Steps 1, 2, and 3 occur if there is no gas on start.

*Power Interruption on Start:*

1. Only the pilot gas flows.
2. Operating sequence resumes when the power is restored.

*Power Interruption During Running Cycle:*

1. The main burner flame is extinguished.
   a. The pilot gas remains lighted.
2. The warp switch cools.
   a. The main gas valve cannot open until about 2 minutes have elapsed after the power is restored.

*Failure to Ignite Pilot Gas:*

1. The same results occur as when there is no gas on start.

*Limit Switch Opens (in the Temperature-control Circuit):*

1. The main burner flame is extinguished.
   a. The main burner gas cannot flow.

*Warp Switch Heater Burns Out:*

1. Step 5 of the normal lighting sequence does not occur. Subsequent steps, therefore, cannot occur.

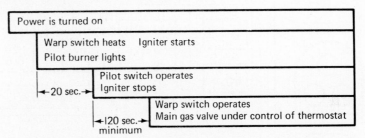

**Figure 10-3.** Steps of normal lighting sequence.

See Figure 10-3 for a schematic of the normal lighting sequence.

The low-voltage electricity is supplied by a Class II transformer and must supply a minimum of 24 volts ac, 60 cycles during the ignition and running cycles.

Repairs of this control cannot be made in the field. These controls, when requiring attention, should be returned to the factory and new ones should be installed according to the manufacturer's specifications.

A typical unit's wiring diagram is shown in Figure 10-4.

## Spark (Electronic) Ignitor

The purpose of an electronic spark ignitor is to light or relight, automatically and instantly, either the pilot gas or the main gas, depending upon the type used.

There are three major types of spark ignitors: (1) those that light the pilot gas with a spark across electrodes, (2) those that light the main burner with a spark across electrodes, and (3) those that light the main burner gas with a spark-plug-type device. See Figure 10-5.

These are solid-state controls that are universal in application and should be given careful consideration when pilot or main burner lighting problems occur.

*Principles of Operation.* These two types of ignition systems mentioned earier are commonly referred to as direct spark ignition (DSI) and intermittent ignition device (IID). The IID system modifies the standing-

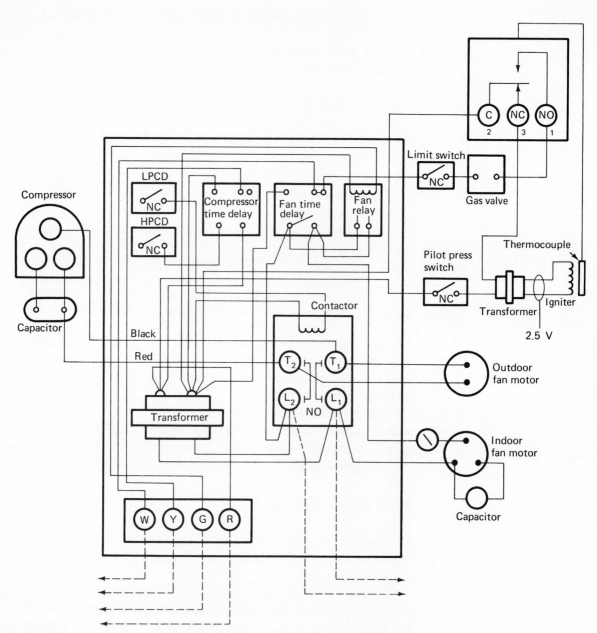

**Figure 10-4.** Typical unit wiring diagram.

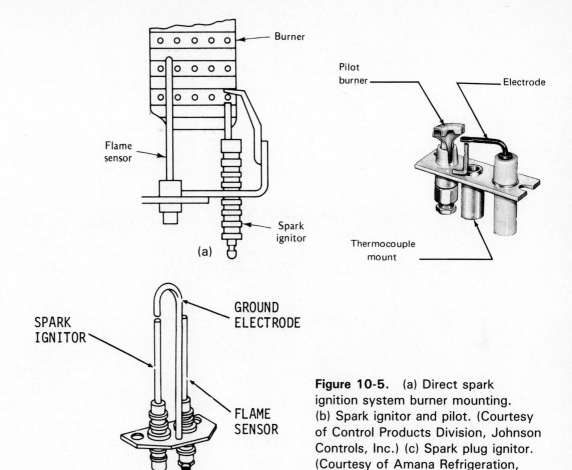

**Figure 10-5.** (a) Direct spark ignition system burner mounting. (b) Spark ignitor and pilot. (Courtesy of Control Products Division, Johnson Controls, Inc.) (c) Spark plug ignitor. (Courtesy of Amana Refrigeration, Inc.)

pilot system in that it uses solid-state electronic circuitry and a flame sensor to replace the thermocouple and the pilot-safety device normally used on standing-pilot-type systems.

*Sensing Methods.*   When the standard thermocouple type system is used, heat is necessary to operate the thermocouple. Heat is not required with the IID system because a method known as *flame conduction,* or *rectification,* is used. In order to understand the principle of flame conduction and rectification, we must first know and understand the structure of a gas flame. See Figure 10-6.

When gas is burning with the proper air-to-gas ratio, there will be a blue flame having three zones:

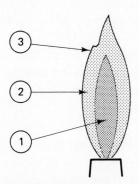

**Figure 10-6.** Gas flame structure. (Courtesy of Carrier Corp.)

*Zone 1:* An *inner cone* that will not burn because there is too much fuel present.

*Zone 2:* A blue envelope, known as the *intermediate cone,* around the inner, fuel-rich cone. In this envelope there is a mixture of vapor from the fuel-rich inner cone and the secondary, or surrounding, air. This is the point at which combustion occurs.

*Zone 3:* A blue envelope known as the *outer cone.* It contains an excessive amount of air.

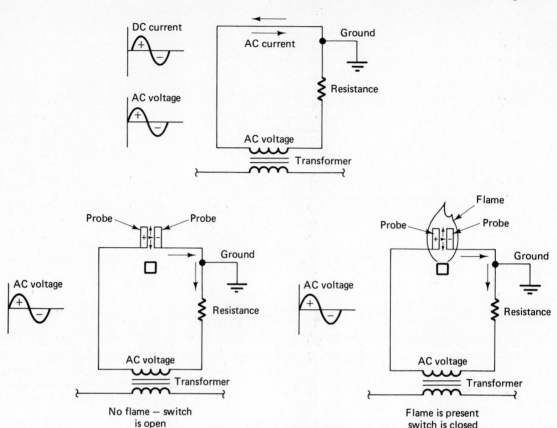

**Figure 10-7.** Electric current through a flame. (Courtesy of Carrier Corp.)

We are primarily interested in the second cone because this is where the combustion occurs. This is the area where the best flame sensing is possible.

A flame is just a series of small, controlled explosions that cause the immediate area to become ionized. Ionization causes the atmosphere to become conductive. This conductive characteristic is what is so important to proper flame conduction. See Figure 10-7.

*Operation:* In operation, the flame can be thought of as a switch. This switch is located between the pilot-burner tip and the flame sensor. When there is no flame between the pilot tip and the sensor, the switch is open. When the flame is in contact with both the pilot tip and the flame sensor, the switch is closed. In Figure 10-7, the flame is used to conduct an ac signal. Both of the probes have approximately the same amount of area exposed to the flame, and the electric current flows through the flame. Unfortunately, this is not good enough to be used as a safety signal because it does not identify the current conducted by the flame. Therefore, a short could mistakenly be identified as a flame. So that the flame can be properly identified, an important difference must be noted. This difference is known as *flame rectification.* See Figure 10-8.

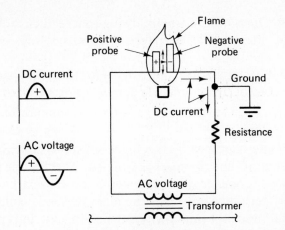

**Figure 10-8.** Flame rectification. (Courtesy of Carrier Corp.)

*Flame Rectification.* In flame rectification, the flame and probes are used in a similar manner but with one major difference. The area of one probe exposed to the flame must be greater than the area of the other probe exposed to the flame.

As can be seen in Figure 10-8, the flame is still used to conduct the ac signal. Both probes are in contact with the flame. The probe having the largest surface area attracts more electrons than the smaller probe. Because of this it becomes the negative probe. The flow of electrons is from the positive probe to the negative probe. Also note that the ac voltage sine wave is not changed, but the negative portion of the current sine wave is no longer shown. The positive portion now represents a pulsating dc electric current. This is the phenomenon of flame rectification.

To make use of this principle, a pilot and flame sensor have replaced the two probes. See Figure 10-9. After ignition of the pilot, a dc microamp current flow is conducted through the flame from the flame to the sensor, (the positive probe), and to the pilot tip (the negative probe). The pilot tip, in this case, acts as the negative probe, completing the circuit to ground. The sensing circuit uses this dc current flow to energize a relay which energizes the main gas valve.

**Figure 10-9.** Pilot and flame as used in flame rectification. (Courtesy of Carrier Corp.)

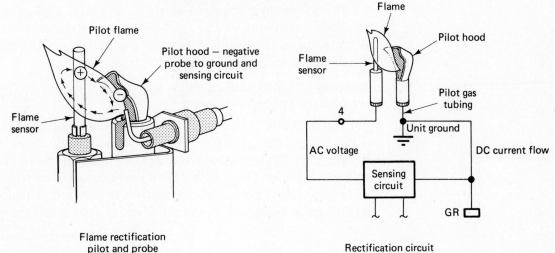

Flame rectification
pilot and probe

Rectification circuit

*Sequence of Operation:*     The following general dicussion describes the sequence of operation of electronic ignition systems:

1. When the temperature control calls for heat, the spark transformer in the control circuit and the pilot valve are energized automatically.
2. The spark ignites the pilot gas on each demand for heat from the thermostat.
3. The flame is proven by the flame sensor. The ignition control is then shut off and the spark is stopped. At this same time, the main gas valve is opened. (Some models allow the spark to continue for a short period of time after the main burner is ignited). On 100% lockout models a shutdown of the entire system is provided if the pilot gas should not light within a fixed amount of time (usually seconds).
4. The main burner gas lights and the system operates normally.
5. When the temperature control is satisfied, the main burner valve and the pilot valves are de-energized, stopping the flow of all gas to the burners.

*Application Guidelines.*     The following conditions have a direct bearing on every IID application.

*Voltage:*     The voltage supplied to electronic ignition controls should be within the ranges listed below:

> 120-volt ac controls: The voltage should be from 102 to 132 volts ac.
> 24-volt ac controls: The voltage should be from 21 to 26.5 volts ac.

All 24-volt ac ignition systems should be equipped with a transformer that will provide sufficient power under maximum load conditions.

*Gas Pressure:*     The gas pressure to the unit should be a minimum of a 1-inch (25.4-millimeter) water column above the equipment manufacturer's recommended manifold pressure. Under no circumstances should the inlet gas pressure be below the equipment manufacturer's recommended inlet gas pressure.

The maximum inlet gas pressure for natural gas applications should be a maximum of a 10.5-inch (266.7-millimeter) water column. On L.P. gas applications the inlet gas pressure should be, at the maximum, a 14-inch (355.6-millimeter) water column.

*Temperature:*     Electronic ignition controls should not be exposed to temperatures above 150°F (66°C) or below −40°F (−40.0°C).

*Pilot Applications:*     The application of the pilot and flame sensor are the most important aspects of the IID application.

The pilot flame must touch the pilot-burner tip and surround the flame-sensor probe. To verify that the proper amount of current is flowing through the pilot flame, a microammeter is required. The amount of current needed is dependent upon the type of system and the manufacturer's design of the system. If the minimum signal is not maintained at all times, the main burner will cycle rapidly or the main flame may not be ignited at all. On systems using flame rectification, the response time may be as little as 0.8 second from the time that the flame is lost. Any deflection of the pilot flame away from the sensor or the tip of the pilot burner could result in rapid cycling of the main burner gas valve or could prevent the main burner from coming on.

Other conditions that can cause failure of ignition of the main burner gas or rapid cycling of the main burner are (1) that the pilot flame is too small or (2) that the gas pressure is too low for proper pilot-flame impingement on the flame sensor. In either of these cases, the pilot gas may ignite, but the main-burner gas valve will not be energized. It is also possible for drafts or unusual air currents to deflect the pilot flame away from the sensor. Deflection of the pilot flame may also be caused by the main-burner gas ignition concussion or main-burner flame rollout.

Another point that should be considered is the condition of the pilot flame. If the pilot flame is hard and blowing, the grounding area of the pilot is reduced to a point where the necessary current is not being maintained, and a shutdown of the system results.

The positioning of the flame sensor is also critical in the pilot application. Positioning of the flame sensor should be such that it will be in contact with the second, or combustion, area of the pilot flame. Passing the flame sensor through the inner cone of the pilot flame is not a recommended procedure. Because of this, a short flame sensor may provide a much better signal than a longer one. The final determination of the sensor location (length) is best determined by the use of a microammeter.

### Spark Plug and Sensor

The following is a description of the spark plug and sensor main-gas ignition systems as used on the Lennox G 14 Series units.

The spark plug and sensor are located on the lower left side of the combustion chamber. See Figure 10-10. Note that the sensor is the top

**Figure 10-10.** Location of spark plug and sensor on Lennox G 14 series furnace. (Courtesy of Lennox Industries, Inc.)

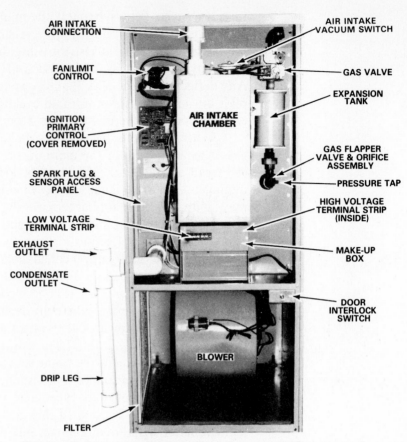

**Figure 10-11.** Location of primary control. (Courtesy of Lennox Industries, Inc.)

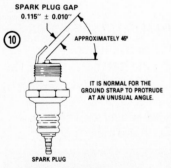

**Figure 10-12.** Spark plus gap setting for Lennox G 14 series furnace. (Courtesy of Lennox Industries, Inc.)

plug; it is longer than the spark plug. The spark plug is below the sensor. These plugs cannot be interchanged because they have different thread diameters.

The spark plug is used in conjunction with a primary control for igniting the initial gas and air mixture. See Figure 10-11. The temperatures in the combustion chamber keep the spark plug free from oxides, and it should not need any regular maintenance. Compression rings are used to form a seal to the combustion chamber.

The spark plug must have a setting within certain specifications for proper operation. See Figure 10-12. Note that in this case a ground-strap angle is used that is unusual when compared to other spark plug ignition systems. A feeler gauge can be used for checking the gap.

The sensor is also a spark plug type with a single center electrode (no ground strap) and compression rings to form a seal to the combustion chamber. It also should not need any regular maintenance.

*Operation.* The following is a description of the sequence of operation of the Lennox spark-ignition system. See Figure 10-13.

1. The line voltage feeds through the door interlock switch. The blower-excess panel must be in place to energize the unit.

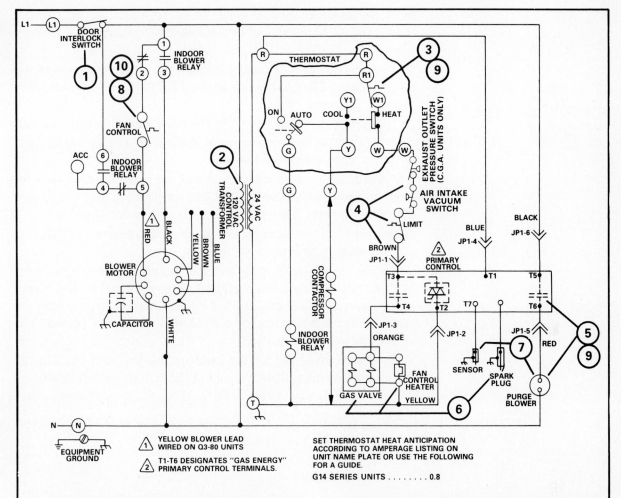

**Figure 10-13.** Sequence of operation schematic for Lennox G 14 series unit. (Courtesy of Lennox Industries, Inc.)

2. The transformer provides the 24 volts ac to the control circuit.
3. When heating is demanded from the thermostat, the thermostat heating bulb contacts are closed.
4. The control circuit voltage passes from the *W* leg through the exhaust outlet pressure switch [C.G.A. (Canadian Gas Association)

only], the air intake vacuum switch [A.G.A. (American Gas Association) and C.G.A. units], and the limit control to energize the primary control.

5.  The electric power then passes through the primary control to the purge blower. The blower is energized for about 30-seconds, providing a prepurge cycle.

6.  At the end of the prepurge cycle, the blower continues to run and the gas valve, the fan control heater, and the spark plug are all energized for approximately 8 seconds.

7.  The sensor determines whether or not the main burner gas is ignited by flame recitification and de-energizes the spark plug and the purge blower. The combustion process is continued.

8.  After about 30 to 45 seconds, the fan-control contacts close to energize the indoor blower motor on low speed.

9.  When the thermostat is satisfied, the heating bulb contacts open. The primary control is de-energized, interrupting electrical power to the main gas valve and the fan control heater. At this time the purge blower is energized for a 30-second post purge period. The indoor blower remains on.

10.  When the recirculating air temperature reaches 90°F (32°C), the fan-control contacts open, stopping the indoor blower motor.

**CONTROLS**   The following is a description of the most popular individual controls used on electronic ignition systems.

### 100% Lockout Module

The *100% Lockout modules* are for use with any non-100% shutoff ignition control. They should be used only according to the manufacturer's recommendations. The purpose of this control is to provide the lockout function needed on this type of system. The lockout function is intended to shut the system down completely if the pilot gas fails to ignite. To initiate a new ignition period the electric power must be interrupted for at least 30 seconds. See Figure 10-14.

**Figure 10-14.**   100% lock out module. (Courtesy of Johnson Controls, Inc., Products Division)

### Spark Ignitor

The spark ignitor consists of an inner electrode made with a ceramic insulator, bracket, and ground strap. Its purpose is to create a spark for direct ignition of the main burner gas. See Figure 10-15. The gap between the electrode and the grounding strap should be set as recommended by the equipment manufacturer. The maximum temperature ratings, as designated by the specific unit manufacturer, should not be exceeded.

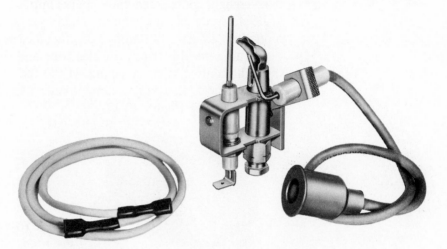

**Figure 10-15.** Direct spark ignitor. (Courtesy of Johnson Controls, Inc., Control Products Division)

### Retrofit Intermittent-Pilot Gas-Burner Ignition Systems

*Retrofit intermittent-pilot gas-burner ignition systems* are complete kits that are made up of the various components required to convert a conventional standing-pilot system to a cycling-pilot control system. See Figure 10-16. These kits are energy-saving devices that can be mounted on furnaces, boilers, and other types of controlled heating systems that have a 24 volt ac control circuit. Otherwise, the control system must be converted to 24 volts ac before the system can be installed.

**Figure 10-16.** Retrofit intermittent-pilot gas-burner ignition system. (Courtesy of White-Rodgers Division, Emerson Electric Co.)

21D18-3

These conversion kits are available for either natural gas or L.P. gas. Be sure to use the correct one for the application. The installation of these kits must be according to manufacturer specifications if satisfactory results are to be realized.

### Pilot-Relight Control

These ignition controls generate spark impulses to light the pilot gas. The relight control generates sparks until a pilot flame is sensed between the electrode and the ground. The flame is detected through flame conduction (ability of a flame to conduct a current). When a flame and current are sensed between the electrode and the pilot-burner ground, the relight control stops sparking. If the flame is extinguished during the call for heat, the relight control begins sparking the instant the flame goes out. See Figure 10-17.

**Figure 10-17.** Pilot-relight control. (Courtesy of White-Rodgers Division, Emerson Electric Co.)

**Figure 10-18.** Flame rectification sensor. (Courtesy of Johnson Controls, Inc., Control Products Division)

The *flame-rectification sensor* is made from an electrical conducting material that is supported by a ceramic insulator and mounting bracket. See Figure 10-18. The purpose of the sensor is to detect the presence of a main burner flame. It is mounted on the main burner so that when the main-burner gas is ignited, the flame will envelop the sensor and permit the electric current to flow through the flame to the ground. These sensors do not require temperature to operate. They require only electrical continuity so that the flame touches both the sensor and the main burner at the same time.

### Ignitor-Sensor

*Ignitor-sensor devices* are generally nonprimary-aerated combination pilot burner and ignitor. They are designed to be used with the controls recommended by the ignitor-sensor manufacturer. See Figure 10-19.

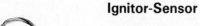

**Figure 10-19.** Ignitor-sensor. (Courtesy of Johnson Controls, Inc., Control Products Division)

### REVIEW QUESTIONS

1. Name the two general types of automatic gas-burner ignition systems.
2. What type burner is the glow coil generally used to light?

3. On glow-coil ignition systems, how long is the purge-time delay if the pilot should go out?

4. Name the three major types of spark ignitors.

5. In what type of ignition system is flame conduction, or rectification, used?

6. Name the three zones of a properly burning flame.

7. When using flame rectification, what is the major difference between the probes?

8. When using flame rectification, which probe will attract more electrons?

9. What part does the flame play in the flame-rectification principle?

10. What is the maximum temperature to which electronic ignition controls should be exposed?

11. What is the purpose of the 100% lockout module?

12. What is the device used to create a spark for direct ignition of the main-burner gas?

13. What is the purpose of the retrofit intermittent-pilot gas-burner ignition system?

14. How is the flame detected in a pilot-relight control system?

15. What is the purpose of the flame-rectification sensor?

# 11 Water Heating and Cooling Controls

Heating and cooling a structure with water are common in large commercial systems and where energy conservation is of prime importance. Control of the boiler and the chiller requires a special type and application of the controls. These controls will be separated in this chapter under two different headings: boiler controls and chiller controls.

**DEFINITIONS**

The following are the definitions of boiler and chiller controls.

**Boiler Control:** A *boiler control* may be defined as any control that provides safe, automatic, and economical operation of a boiler.

**Chiller Controls:** A *chiller control* may be defined as any control that provides safe, automatic, and economical operation of a chiller.

**WATER-LEVEL CONTROL**

The purpose of a *water-level control* is to stop the burners in the event of low water in a boiler in order to prevent damage to the boiler. The water-level control is a float-operated control, which senses the water level inside the boiler and will cut off the fuel to the main burner when a predetermined level is reached. See Figure 11-1. This control is designed to operate on boilers with pressures up to approximately 30 pounds per

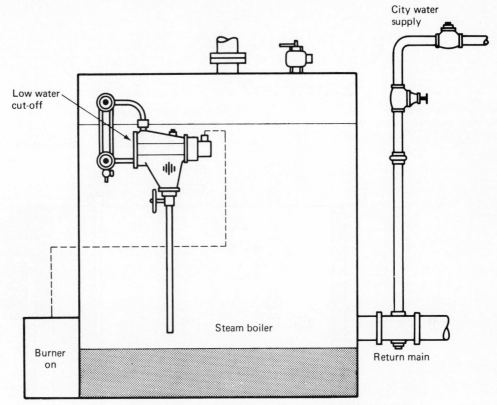

**Figure 11-1.**  Float-operated low-water cutoff.

square inch gauge (307.98 kilopascals). It is mounted on the boiler so that the switch will interrupt the control circuit to the main fuel valve when the boiler water level drops to a dangerously low level.

The low-water cut-off control is mounted on the boiler at the desired water level. See Figure 11-2. There are several locations and piping arrangements available for this type of installation. Because there is no normal water line to be maintained in a hot-water boiler, any location of the control above the lowest permissible water level is satisfactory. A steam boiler does, however, have a specific water level that must be maintained, and the recommendations of the boiler manufacturer should be followed.

The construction of a hot-water boiler is essentially the same as a steam boiler. The major difference is in the way the steam boiler is operated. Most of the conditions causing low water to occur in a steam boiler will also hold true for a hot-water boiler.

*Operation.*   The water line in the boiler and the water line in the control drop at the same time and are at the same level. This lowering of the water line in the control-float chamber causes the float to drop.

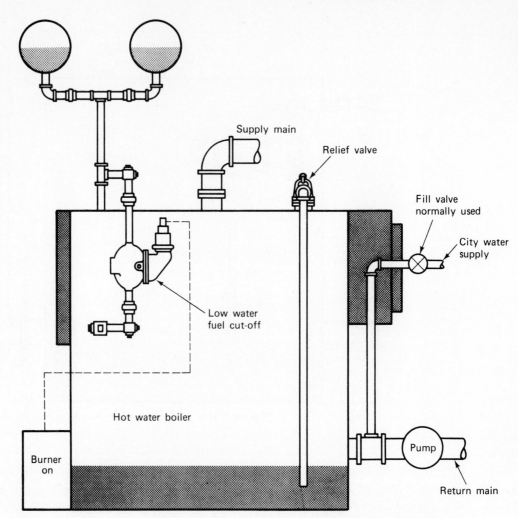

**Figure 11-2.** Installation of low-water cutoff.

The action of the float causes the control-switch contacts to open and interrupt the control circuit to the automatic burner. See Figure 11-3.

This action provides us with a basic safety control for boilers. It is a means of stopping the automatic firing device if the water level should drop below the minimum safe level.

The following statement appears in a booklet, "Recommended Practices for Installation," published by a leading utility company (McDonnell Basic Safety Controls for Hot Water Space Heating Boilers, Bulletin No. P-30C, Chicago, Ill., 1962 p. 6). Their experience in the heating field prompts them to write:

A low water cut-off which will cut off the fuel supply before the water level reaches a low danger point, or a water feeding device with a shut-off, shall be attached to all steam and hot water boilers.

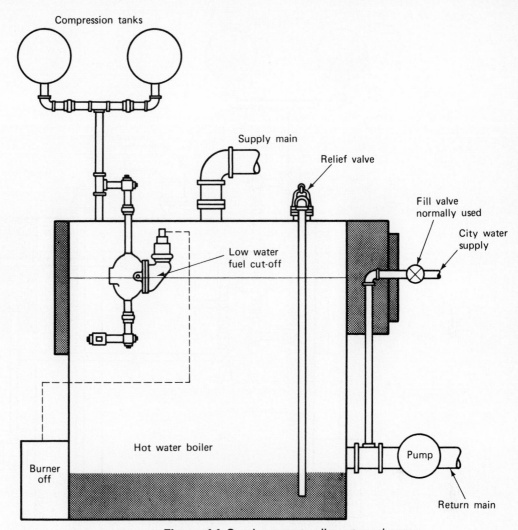

**Figure 11-3.**   Low-water line-stops burner.

### Combination Feeder and Low-Water Cutoff

If the low-water cutoff could be absolutely relied upon to stop the automatic burner each time a low-water condition occurred, then the problem would be solved completely. However, experience has proved that under certain circumstances, the low-water cutoff cannot fulfill its duties.

The combination feeder and low-water cutoff offers much more safety than the low-water cutoff alone can. It covers almost all installations and provides the most complete measure of safety. See Figure 11-4.

This control provides

1. The mechanical operation of feeding water to the boiler as fast as it is discharged through the relief valve.

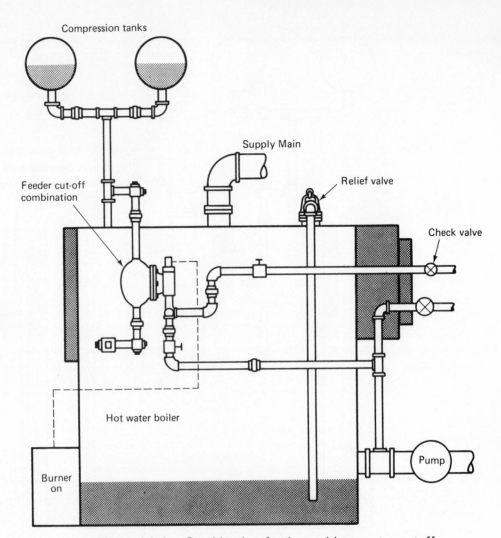

**Figure 11-4.** Combination feeder and low-water cutoff.

2. The electrical operation of stopping the burner when a low-water situation occurs.

This combination of water feeding and control of the electrical circuit offers the best safeguards, which are the best and most complete recommendation for boilers.

*Operation.* In operation, the combination feeder and low-water cutoff control admits water into the boiler to maintain the desired water level in the boiler, while at the same time completing the electrical power to the control circuit that operates the burner in response to the requirements of the boiler. Should the water leave the boiler faster than the feeder can admit it, the low-water cutoff stops the main burner before the boiler becomes overheated and perhaps ruined.

**SWITCHES**  The low-water cutoff and the combination feeder and low-water cutoff control have electrical contacts of various configurations. See Figure 11-5. These contacts are directly connected to and are operated by the position of the water float in the control-float chamber.

In Figure 11-5(b) the NC contacts complete the control circuit and allow the burner to operate on demand from the temperature or pressure control.

In Figure 11-5(c) the diagram shows that when a low water level condition occurs, the control circuit is opened and the NO alarm contacts are made, sending a danger signal to the operating engineer.

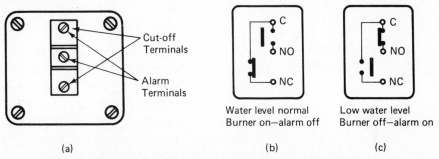

Figure 11-5. (a) Switch terminal locations. (b) Water level normal, burner on-alarm off. (c) Low-water level, burner off-alarm on.

**BOILER TEMPERATURE CONTROL (AQUASTAT)**  In normal operation, the boiler temperature control regulates the operation of the main burner by interrupting the control circuit. The sensing element is installed in the boiler through openings provided by the boiler manufacturer. See Figure 11-6. The switch may be either an SPST snap-acting switch or a mercury tube. Both of these switches are actuated by a helically wound bimetal. These controls are adjustable so that the proper operating temperature of the unit can be selected.

Figure 11-6. Boiler temperature control. (Aquastat) (Courtesy of White-Rodgers Division, Emerson Electric Co.)

*Operation.*  As the water temperature inside the boiler drops to the cut-in temperature, the sensing element moves to a position that causes the switch to close and complete the control circuit to the main burner gas valve. See Figure 11-7. As the water is heated to the cut-out temperature setting of the aquastat, the switch contacts open and interrupt the control circuit, de-energizing the main burner gas valve.

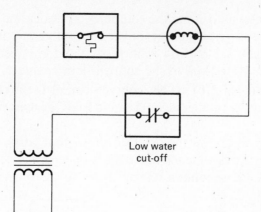

**Figure 11-7.** Basic boiler wiring diagram with temperature control.

**BOILER PRESSURE CONTROL** The *boiler pressure control* is used on steam boilers and is mounted above the boiler proper to sense the pressure at the most critical point in the system. See Figure 11-8. The switch is actuated by a diaphragm and bellows, which expand with an increase in pressure and contract with a decrease in pressure. The contacts in these controls are wired in series in the control circuit and either make or break the control circuit in response to the pressure inside the system. See Figure 11-9. Boiler pressure controls have an adjustable scale so the proper operating pressure for the system can be maintained. See Figure 11-10.

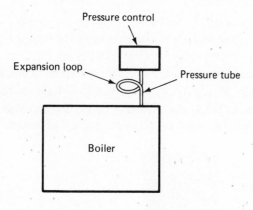

**Figure 11-8.** Boiler pressure control installation.

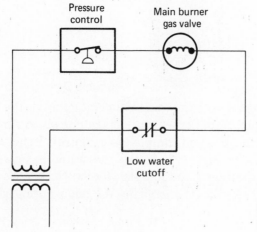

**Figure 11-9.** Basic boiler wiring diagram with pressure control.

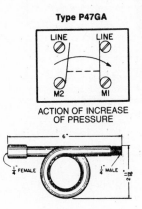

**Type P47GA**

ACTION OF INCREASE
OF PRESSURE

**Figure 11-10.** Boiler pressure control. (Courtesy of Johnson Controls, Inc., Control Products Division)

*Operation.* As the pressure inside the steam system drops to the cut-in setting of the control, the switch contacts close, completing the control circuit to the main-burner gas valve. As the burner operates, the pressure inside the system is increased to the cut-out setting of the pressure control. At this pressure the contacts open, interrupting the control circuit and de-energizing the main-burner gas valve.

**HIGH-LIMIT CONTROL**  The *high-limit control* is a safety device that is wired into the control circuit to prevent overheating the boiler in case the temperature or pressure control fails to function properly. See Figure 11-11. These controls may use either SPST or SPDT switches. The SPST type only interrupts the control circuit and stops operation of the main-burner gas valve. The SPDT models may be used for several different functions, such as energizing a circulator pump or energizing an alarm circuit to alert the user that a problem has occured. See Figure 11-12. The high-limit control is wired into the control circuit in series with the other controls. See Figure 11-13.

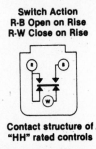

**Switch Action**
**R-B Open on Rise**
**R-W Close on Rise**

**Contact structure of**
**"HH" rated controls**

**Figure 11-11.** High-limit control. (Courtesy of White-Rodgers Division, Emerson Electric Co.)

**Figure 11-12.** Contact configuration of high-limit control. (Courtesy of White-Rodgers Division, Emerson Electric Co.)

*Operation.* If the boiler temperature rises above the setting of the pressure or temperature controller, the high-limit control will interrupt the control circuit, de-energizing the main-burner gas valve to prevent

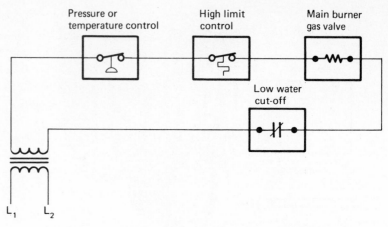

**Figure 11-13.**    Basic boiler wiring diagram.

overheating the boiler. The secondary circuit, if used, will then be energized to perform the desired function, such as starting the circulator pump, or energizing the alarm circuit. As the temperature or pressure of the boiler drops to the cut-in setting of the control, the contacts close completing the control circuit. The boiler is again started for normal operation.

**Figure 11-14.**    Cutaway view of high-low fire valve.

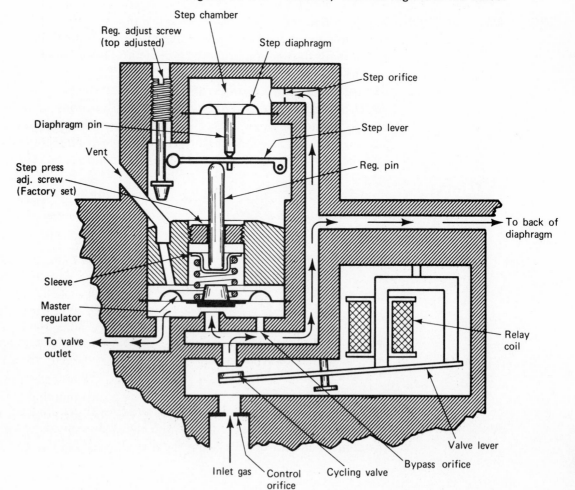

**HIGH-LOW FIRE VALVES**

The *high-low-off* type of combination diaphragm valve provides all the manual and automatic control functions required for operation of gas-fired heating equipment. These valves have an internally vented diaphragm-type main control valve and also a separate thermomagnetic safety valve with a pilot-gas adjustment and a pilot filter. Most models include a pressure-regulator function for use with natural gas. See Figure 11-14.

The main-line gas-valve diaphragm opens and closes in response to an aquastat or a boiler-pressure control. The main-line gas-diaphragm valve functions simultaneously as a pressure regulator on natural gas models. The L.P. gas models are not equipped with pressure regulators.

When the electrical circuit is completed between the C and $W_1$ terminals of the valve, the valve automatically opens to the preset low-fire position. Then, when the circuit is completed between C and $W_2$, the valve opens to the high-fire position. The shift from low-fire to high-fire is accomplished by a heat motor in the valve operator. Time must be allowed for this heating action to be completed. See Figure 11-15.

This type of valve is used when the full BTU rating of the boiler is not needed but some heating is required.

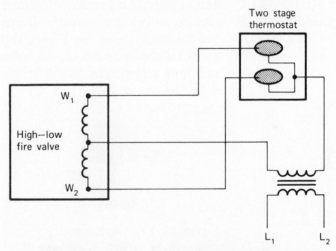

**Figure 11-15.** High-low fire wiring diagram.

**LIQUID-LEVEL FLOAT SWITCH**

*Liquid-level float switches* are designed with several purposes in mind. They are mounted so that they can sense the liquid level in a tank. See Figure 11-16. The switch may be either an SPST or an SPDT type. They can be wired to open one circuit and close another circuit when the liquid level rises above or falls below the required level. When the liquid level reaches a predetermined level, the float will open one set of contacts and close another, depending upon the design of the control. They are used for monitoring steam-boiler condensate tanks, waste-water tanks, and any other type tank requiring automatic filling and emptying of the fluid in it. They are wired so that they will control the operation of a pump motor used to pump out the tank at the proper time or a fill-solenoid valve that fills the tank with fluid when required.

**Figure 11-16.** Liquid-level float switch. (Courtesy of Johnson controls, Inc. Control Products Division)

*Operation.* When the float reaches the point of making either set of contacts, the contacts complete the circuit to the desired function and either start a pump motor or energize a solenoid cell. The fluid is either removed from the tank or the tank is filled to the desired level. The float then opens that set of contacts to stop the function. In some cases the extra set of contacts may be used to energize an alarm system or an information light to indicate that the desired function is either complete or needs attention. Again, the use of these switches is limited only by the imagination of the user.

**AUTOMATIC CHANGEOVER CONTROL**

The *automatic changeover control* is designed to provide automatic summer-winter changeover of the thermostat action in hydronic heating and cooling systems. See Figure 11-17. The control is available in either the direct clamp-on type or the bulb clamp-on type. They have an SPDT switching action. The switch opens red to black at 85°F (29.4°C) and red to brown at 70°F (21.1°C).

**Figure 11-17.** Automatic changeover switch. (Courtesy of White-Rodgers Division, Emerson Electric Co.)

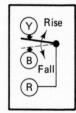

**Figure 11-18.** Switching action of automatic changeover control.

*Operation.* When the boiler lights and the water temperature inside the pipe rises to 89°F (32°C) at the point where the control is attached, the switching action closes the red to yellow circuit, which in turn changes the function of the thermostat to the heating mode. The controls will function in the heating mode until the boiler is shut down and the chiller is started. When the water temperature inside the pipe at the point where the control is attached drops to 62°F (17°C), the switch opens the circuit between the red and yellow wires and makes the circuit between the red and blue or the red and black wires. The thermostat is changed over into the cooling mode and functions to provide the required cooling control of the space. See Figure 11-18.

**CHILLER PROTECTOR**    *Chiller protector* controls are designed specifically to protect chillers from freezing and perhaps damaging the tubes. See Figure 11-19. When the temperature drops to the predetermined cut-out setting of the control, the contacts open the control circuit and stop the compressor. The switch is an SPST type that opens on a fall in temperature. While the compressor is off, the chiller de-ices so that the unit will function properly when it is restarted. The cut-out point of these controls is usually about 38 °F to 40 °F (3.3 °C to 4.4 °C). The sensing element is generally mounted in the water line leaving the chiller so that the water is sensed after it has passed through the chiller. Chiller protectors may be either automatic-reset or manual-reset. These controls are safety devices and should never be taken out of the circuit. If they are tripping, the problem must be found and corrected.

**Figure 11-19.**   Chiller protector. (Courtesy of Robertshaw Controls Company, Uni-Line Division)

*Operation.*    When the chiller is operating and the water is being cooled, the system will function as it should. If for some reason the water temperature reaches the cut-out temperature of the chiller-protector control, the control contacts open and interrupt the control circuit to the compressor starter. The compressor stops and the chiller temperature rises to the cut-in point of the control. If the automatic-reset type is used, the compressor will again start operating and the system will function as it was designed. If the control is a manual-reset type, the operator must reset the control before compressor operation can be resumed.

**FREEZESTAT/LOW-**    The *freezestat/low-limit control* is designed to protect heating and/or
**LIMIT CONTROL**    cooling coils or other similar devices from freeze damage. See Figure 11-20. This control responds to the lowest temperature sensed along its entire bulb length. It has an SPST switching action that opens the contacts on a fall in temperature. They may be either manual- or automatic-reset types, depending upon the model used. These controls should be set with a cut-out point that will prevent freezing of the coil or protected device, usually above 32 °F (0 °C).

**Figure 11-20.** Freezestat/low-limit control. (Courtesy of Robertshaw Controls Company, Uni-Line Division)

*Operation.* While the compressor is operating and the air leaving the coil is being cooled, the unit will operate as designed. However, if something should happen to reduce the air flow or some other control should fail to function as it was designed, the freezestat/low-limit control contacts will open, interrupting the control circuit and stopping the compressor. The compressor remains off until the control either automatically resets or is manually reset. When the cut-in setting of the control is reached, the compressor again starts and the system functions as designed, unless something is wrong with either the indoor-air delivery or the refrigerant charge in the system.

**FLOW SWITCH**    *Flow switches* are used in hydronic systems to protect the boiler and chiller from damage in case the pump fails to operate or fails to provide the proper flow of water through the system. See Figure 11-21. The switch has SPDT contacts, which are used to make one circuit while breaking another circuit. See Figure 11-22.

On an increase in the flow of water, the contacts make from red to yellow. On a decrease in the water flow, the contacts make from red to blue. This switching action can be used to stop the compressor or burner, energize an alarm circuit, start another pump, or do any combination of other uses that will provide the desired function of the unit.

**Figure 11-21.** Flow switch. (Courtesy of Johnson Controls, Inc., Control Products Division)

**Figure 11-22.** Flow switch switching action. (Courtesy of Johnson Controls, Inc., Control Products Division)

*Operation.*    While the proper amount of water is flowing through the system, the contacts are made from red to yellow and the compressor or boiler may operate as desired. However, if the pump should fail, the contacts will change red to blue, energizing the proper circuit to provide the function desired. When the water flow is again resumed, the flow switch contacts change to make red to yellow, and the unit resumes normal operation.

**FAN CENTER**    *Fan centers* provide low-voltage control of line voltage fan, pump motors, and auxiliary circuits in forced warm-air or hydronic heating, cooling, or heating-cooling systems. See Figure 11-23.

These controls have external wiring terminals for convenience in making the electrical connections. There is a relay enclosed that is used to control the desired piece of equipment. A transformer is included to provide the low voltage for the control circuit. There are two sets of contacts: a main pole and an auxiliary pole. The voltage and amperage of each of these poles must not be exceeded or damage will be likely. The wiring should conform to the diagram that accompanies the control.

**Figure 11-23.**  Fan center. (Courtesy of White-Rodgers Division, Emerson Electric Co.)

## REVIEW QUESTIONS

1.  Define a boiler control.
2.  Define a chiller control.
3.  What is the purpose of a water-level control?
4.  Which boiler control offers safeguards that are the best and most recommended for boilers?
5.  Which boiler-control controls both the feedwater and the electrical circuit to the boiler?
6.  What does the boiler temperature control regulate?
7.  Where should boiler pressure controls be located?
8.  What is the primary function of the high-limit control?
9.  Do high-low fire gas valves have regulators in both natural gas and L.P. gas models?
10.  What is the purpose of the liquid-level float switch?

11. What control is designed to provide automatic summer-winter changeover of the thermostat action in hydronic heating and cooling systems?

12. Which control is designed specifically to protect chillers from freezing?

13. What is the purpose of the freezestat/low-limit control?

14. In case of pump failure, what control protects the chiller or boiler from damage?

15. What type of control has both the relay and the transformer included?

# 12

# Oil-Burner Controls

Fuel oil used for heating purposes is capturing its share of the heating market. There are approximately 65 million users of oil heating equipment. Because of the popularity of fuel oil, it is necessary to include controls for safe, automatic operation in a book of this type. Fuel-oil burners use a few controls that are common to gas or electric equipment, such as fan control, limit control, thermostat, and so on.

**DEFINITION**  An *oil burner* is a device that supplies a mixture of atomized fuel oil and air, under pressure, to the combustion chamber, where the majority of combustion takes place.

**PROTECTORELAY (STACK MOUNTED)**  The *stack-mounted protectorelay control* is simply a combustion thermostat used to sense changes in the temperature of the flue gases. These changes in flue-gas temperature operate the stack contacts of the stack control; through the action of the thermal element, they control the operation of the safety switch in the relay unit, assuring safe starting and proper burner operation. Both recycling and nonrecycling combinations of stack controls are available. These units are mounted directly on the flue with the sensing element inside the flue pipe. They should be mounted as near as possible to the furnace or burner and at least 2 feet from the draft regulator. See Figure 12-1.

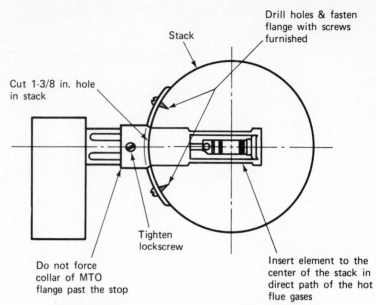

**Figure 12-1.** Typical stack-control mounting.

*Operation.* To start the burner, first be sure that the combustion chamber is free of oil.

1. To put the contacts in step, pull the drive-shaft lever outward $\frac{1}{4}$ inch and release slowly.
2. Move the red reset lever to the right and release.
3. Open the hand valve in the oil-supply line.
4. Set the thermostat to call for heat.
5. Close the electrical switch. The burner should start.

*Ignition Timing:* If the relay drops out too soon after the burner starts, adjust the ignition timing lever toward *maximum*.

*Scavenger Timing:* The following is a description of the scavenger-timing operation:

1. With the burner on, the drive shaft carries the clutch finger outward. The stop arm halts the clutch finger, but the drive shaft moves a small amount farther. This override is necessary for proper sequencing. *Note:* If the clutch finger does not reach the stop arm with the recycle lever at the minimum setting, the bimetal is not getting enough heat. The stack control must be relocated.
2. Allow the burner to run a few minutes; then open and close the electrical line switch. The burner should stop at once.
3. The burner should start again in about 1 minute.
4. If the burner starts too soon, proceed as follows:
   a. Open the line switch; wait 5 minutes for cooling.
   b. Move the cycle lever outward one notch. Close the line switch to start the burner and repeat Steps 1, 2, and 3.
   c. Repeat Steps a and b until the timing is satisfactory.

To check the stack switch, use the following procedure to verify the safety features:

1. Flame failure:
   a. Test for recycling by shutting off the oil-supply hand valve while the burner is operating normally. Restore the oil supply after the burner shuts off. After a short scavenger period, the stack control restarts the burner.
   b. Test for safety shutoff after the flame failure by shutting off the oil-supply hand valve while the burner is operating normally. When the burner shuts off, do not restore the oil supply at this time. The stack control will attempt to restart the system after a scavenger period; then in approximately 30 seconds, the safety switch will lock out. Reset the safety switch, and the burner will restart.

2. Ignition Failure:
   a. Test by turning off the oil supply while the burner is off. Run through the starting procedure, omitting Step 3. A lockout will occur. Reset the red safety switch.

3. Power Failure:
   a. Turn off the power supply while the burner is on. When the burner stops, restore the power and the burner will restart after a scavenger period.

For a typical wiring diagram of an oil burner system, see Figure 12-2.

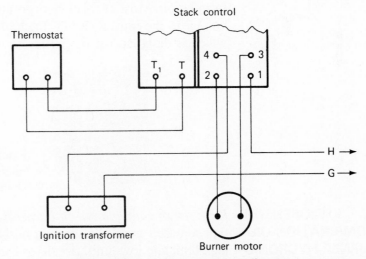

**Figure 12-2.** Typical wiring diagram.

**BURNER-MOUNTED COMBUSTION THERMOSTAT**

Another type of combustion thermostat is mounted on the burner. This thermostat is a hermetically sealed switch whose contacts open and close in response to temperature changes produced by the radiant heat of an oil-burner flame. It is not a light-sensitive device and is not affected by normal soot deposits. Its small size permits easy mounting on the blast tube of the burner, where it will quickly sense the radiant heat of the flame pattern.

*Operation.* This control is designed to monitor the burner flame. In the starting operation the contacts must be in the closed position. The control contacts remain in the closed position until there is enough radiant heat from the flame to cause them to open.

When the proper flame is established, a rise in temperature of the sensing face opens the contacts of the control within seconds and de-energizes a safety switch heater, allowing a normal start and run cycle. In the event of a loss of flame during a burning cycle, the resultant drop in temperature closes the contacts and de-energizes the safety circuit of the relay.

After a normal burner run the combustion-control contacts return to the closed, or starting, position. If an immediate start is called for by the thermostat, an enforced purge period will occur while the combustion contacts and ignition switch are being restored to the closed, or starting, position.

**KWIK-SENSOR CAD-CELL RELAY (PROTECTORELAY) BURNER CONTROLS**

*Kwik-sensor cad-cell relay burner controls* are available for use on either intermittent or interrupted operation of oil burners. They are designed to provide control of the ignition process of oil burner systems. See Figure 12-3. Cad-cell relays are designed to be used in combination with certain types of controls. Be sure that the proper combination is being used for proper operation. The manufacturer provides the proper supporting controls for the type being used. These controls use a solid-state flame-sensing circuit. They can be manually reset after a shutdown of the system. There are low-voltage terminals to simplify the wiring process during installation. Most of them incorporate a transformer to provide the low voltage to the control circuit. Each of them has its own wiring diagram, which is designed for its particular use in the system.

**Figure 12-3.** Kwik-sensor cad-cell relay (Courtesy of White-Rodgers Division, Emerson Electric Co.).

**KWIK-SENSOR COMBINATION OIL-BURNER-HYDRONIC CONTROL**

*Kwik-sensor combination oil-burner-hydronic controls* use an immersion-type aquastat controller and an oil-burner primary control to provide high-limit and low-limit/circulator control for oil-fired hydronic heating systems. See Figure 12-4. They provide intermittent (formerly called constant) ignition of the fuel oil. This type of control is designed to be used with specific supporting controls, and the manufacturer must be consulted for the proper types of controls to be used. Some models are designed to mount directly on the burner and others are designed to mount externally. These are equipped with an armored capillary having a remote sensor. Almost all of them are capable of multiple-zone control with use of the proper valves.

**Figure 12-4.** Kwik-sensor combination oil burner-hydronic control (Courtesy of White-Rodgers Division, Emerson Electric Co.).

## CADMIUM SULPHIDE FLAME DETECTOR

**Figure 12-5.** Flame detector (Courtesy of White-Rodgers Division, Emerson Electric Co.).

*Cadmium sulphide flame detectors* are photoconductive flame-sensing devices that are used for sequencing oil-burner systems. These detectors are mounted so they can sense the oil-burner flame and signal the combination oil-burner and hydronic control on flame detection or failure. See Figure 12-5. These flame detectors are glass-to-metal hermetically sealed plug-in cells to prevent deterioration by humidity, soot, or oil fumes. They are equipped with NEC Class 1 lead wires.

*Operation.* On flame failure, the light-sensitive cadmium sulphide cell, in conjunction with the flame-sensing circuitry, causes the combination oil-burner-hydronic control to shut down the main oil burner. These controls have provisions to bypass the cadmium cell upon reignition of the oil burner for a given period of time until the flame is sensed by the flame detector. Then the oil burner operates in its normal mode.

## MAGNETIC VALVES

*Magnetic valves* are designed for the on-off control of the flow of oil to domestic oil burning equipment. See Figure 12-6. These valves are designed to control the flow of fuel oil no heavier than number 2 with a temperature of 125 °F (52 °C) and with an ambient temperature range of 32 °F (0 °C) to 115 °F (46 °C). Upon interruption of electrical power, the valve closes immediately.

**Figure 12-6.** Magnetic valve (Courtesy of White-Rodgers Division, Emerson Electric Co.).

*Operation.* When the thermostat calls for heat, the burner motor is started and the magnetic valve is energized at the same time. When the thermostat is satisfied, the valve closes immediately. This prevents oil flowing into the firebox; such oil would be ignited on the next on-cycle, causing possible overheating of the combustion area.

## BURNER SAFETY CONTROLS

These *burner safety controls* are the same safety controls that are used on warm-air heating systems and steam or hot-water boilers. They will stop the burner in the case of abnormal pressures or temperatures resulting

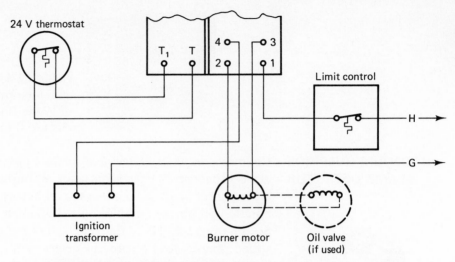

**Figure 12-7.** Wiring diagram with burner safety control.

from insufficient water or air flow. They are wired into the system in the same manner.

A typical wiring diagram is shown in Figure 12-7.

## REVIEW QUESTIONS

1. Define an oil burner.
2. What is a protectorelay?
3. Where should a protectorelay be mounted?
4. To what does the burner-mounted combustion thermostat react?
5. What does the burner-mounted combustion thermostat monitor?
6. What type of oil-burner control can be used on either intermittent- or interrupted-operation oil burners?
7. What type of flame-sensing circuit is used on cad-cell relays?
8. What type of control provides intermittent ignition of the fuel oil?
9. Which control is used for sequencing oil burner systems?
10. What is the purpose of magnetic valves on oil-burning equipment?

# 13

# Head- (Discharge-) Pressure Controls

Head-, or discharge-, pressure controls are necessary when refrigeration and air conditioning equipment must be operated during mild or cold weather conditions. There are two reasons for this: (1) When the head pressure rises above the normal operating point, the economy and performance of the system are drastically reduced, and (2) when the head pressure drops below the normal operating point, there is not enough pressure drop across the flow control device to provide proper distribution and evaporation of the refrigerant in the evaporator.

**DEFINITION**    A *head-pressure control* is any device that maintains the desired head pressure inside the refrigeration system by controlling the amount of coolant delivered to the condenser.

**COOLING TOWER-FAN CONTROL:**    *Tower-fan controls* are devices that are designed to cycle the tower fan to maintain the water temperature at a temperature that will produce the desired head pressure for the refrigeration system. See Figure 13-1. These are single-stage temperature controls that may be equipped with either DPST or SPST contact arrangements. The contacts for the fan occur on a temperature rise. The extra contacts may be used to energize an alarm circuit or to energize some type of emergency equipment to prevent overheating the tower water. The thermostat is wired into the tower-fan circuit, as shown in Figure 13-2.

**Figure 13-1.** Cooling tower fan thermostat (Courtesy of Johnson Controls, Inc., Control Products Division).

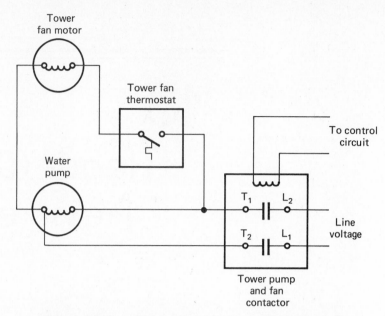

**Figure 13-2.** Wiring diagram for cooling tower fan thermostat.

*Operation:* When the ambient temperature around the tower drops to a predetermined point, the thermostat will sense that the tower water is getting too cold, and the contacts will open and stop operation of the fan. The fan will remain off until the tower water temperature has risen to the cut-in setting of the thermostat. The contacts will then close, starting the tower fan to bring the water temperature down. This type of operation continues until the ambient temperature has risen to the point that constant operation of the tower fan is needed.

**TWO-SPEED CONDENSER-FAN CONTROLS**

Some condensing units are equipped with two-speed fans, which are used for controlling the head pressure. The speeds are controlled by a two-speed condenser-fan thermostat or pressure control. See Figure 13-3. This thermostat is fastened to one of the return bends in the condenser coil.

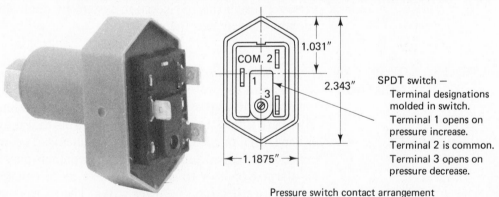

SPDT switch —
Terminal designations molded in switch.
Terminal 1 opens on pressure increase.
Terminal 2 is common.
Terminal 3 opens on pressure decrease.

Pressure switch contact arrangement

**Figure 13-3.** Two-speed condenser fan pressure control (Courtesy of Robertshaw Controls Company, Uni-Line Division).

The pressure control is in the discharge line. See Figure 13-4. The exact location is usually determined by the equipment manufacturer, and these recommendations should be followed for best results. The thermostat changes the fan motor from high to low speed and back to high speed in response to the refrigerant temperature inside the condensing coil. It is a nonadjustable switch that changes the fan motor speed at temperatures recommended by the manufacturer.

*Operation.* When the thermostat inside the building demands cooling, the compressor and the condenser fan are energized. The fan operates in the low speed until the refrigerant temperature has risen to the changeover setting of the thermostat. The fan then changes into high speed at this changeover point. The fan continues to operate in the high-speed mode until the refrigerant temperature drops to the low-speed changeover temperature. Then the fan operates in the low-speed mode.

**Figure 13-4.** Location of condenser fan thermostat (Courtesy of Lennox Industries, Inc.).

When the thermostat inside the building is satisfied, both the compressor and the condenser fan stop. The condenser fan thermostat cools to the ambient temperature and switches to the low-speed position for the next start period.

A typical wiring diagram of a two-speed condenser fan thermostat is shown in Figure 13-5. *(See page 176).*

**LOW-AMBIENT KIT**   The purpose of the *low-ambient kit* is to cycle the condenser fan to maintain the desired head pressure in the refrigeration system. See Figure 13-6. Under certain outdoor conditions the condenser fan should not operate because it will drop the head pressure below the desired pressure for

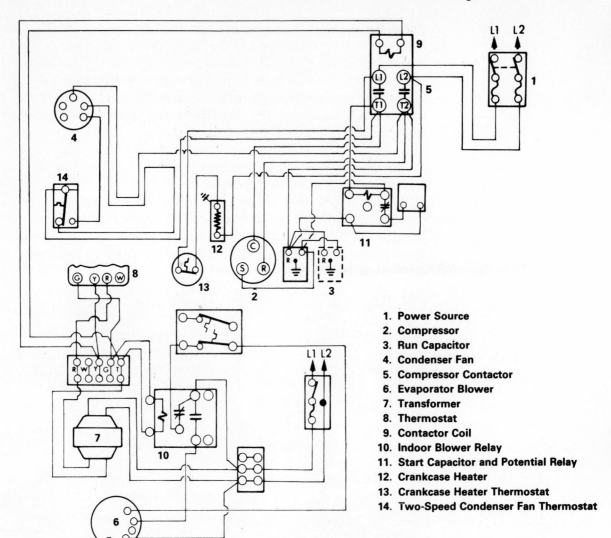

1. **Power Source**
2. **Compressor**
3. **Run Capacitor**
4. **Condenser Fan**
5. **Compressor Contactor**
6. **Evaporator Blower**
7. **Transformer**
8. **Thermostat**
9. **Contactor Coil**
10. **Indoor Blower Relay**
11. **Start Capacitor and Potential Relay**
12. **Crankcase Heater**
13. **Crankcase Heater Thermostat**
14. **Two-Speed Condenser Fan Thermostat**

**Figure 13-5.**  Two-speed condenser fan wiring diagram (Courtesy of Lennox Industries, Inc.).

**Figure 13-6.**  Typical low-ambient kit (Courtesy of Lennox Industries, Inc.).

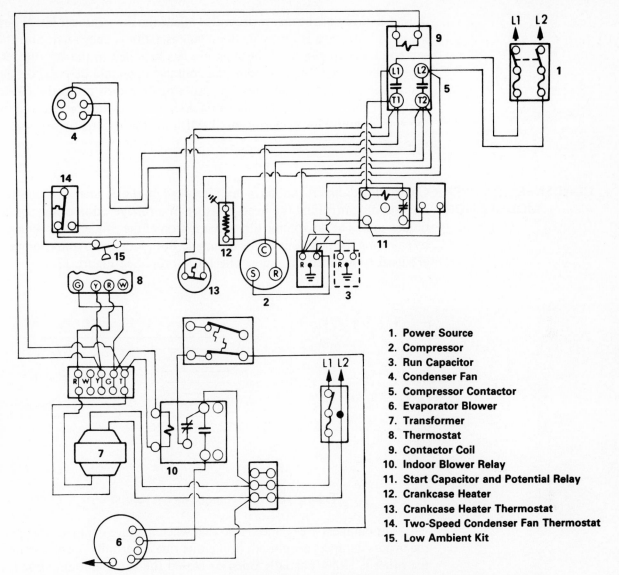

**Figure 13-7.** Typical low-ambient kit wiring diagram (Courtesy of Lennox Industries, Inc.).

1. Power Source
2. Compressor
3. Run Capacitor
4. Condenser Fan
5. Compressor Contactor
6. Evaporator Blower
7. Transformer
8. Thermostat
9. Contactor Coil
10. Indoor Blower Relay
11. Start Capacitor and Potential Relay
12. Crankcase Heater
13. Crankcase Heater Thermostat
14. Two-Speed Condenser Fan Thermostat
15. Low Ambient Kit

proper operation of the unit. There is a high-pressure control furnished with the kit that is wired in series with the condenser fan motor. See Figure 13-7.

The pressure switch cycles the fan motor while allowing the compressor to continue to operate. In most cases when the outdoor ambient temperature drops to the point that the head pressure falls to 140 pounds per square inch gauge (1067.67 kilopascals), the pressure switch contacts open and de-energize the condenser fan motor. When the head pressure builds back up to the cut-in setting of the pressure control, the condenser fan is again started to cool the condenser.

*Operation.* When the thermostat inside the building demands cooling, the compressor and condenser fan are energized at the same time. The condenser fan remains off until the head pressure has increased to the cut-in point of the pressure control inside the low-ambient kit. When the cut-in pressure is reached, the condenser fan is energized, and the fan continues to run until the pressure has dropped to the cut-out setting of the control. At this point the contacts open and de-energize the condenser fan motor. The head pressure again builds up to the cut-in setting of the control, when the condenser fan is again started to cool the refrigerant. This cycle repeats itself until the indoor thermostat is satisfied.

**CONDENSER-DAMPER MODULATION**   Controlling dampers that have been installed on the condenser is another method of controlling the head pressure of a refrigeration or air conditioning unit. The air volume through a section of the condenser is changed by changing the positions of the dampers. Both face and bypass dampers are used to avoid overloading the fan motor. See Figure 13-8.

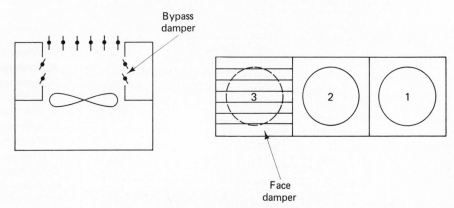

**Figure 13-8.**   Damper control location.

The dampers are connected so that when the face dampers are open, the bypass dampers are closed, or if either one is partially open or closed, the other is either partially open or closed the same amount. For full cooling, the face dampers are completely open and the bypass dampers are completely closed. See Figure 13-9.

These dampers are usually controlled by a pressure switch, which responds to the head pressure of the system being controlled.

*Operation.*   When the indoor thermostat demands cooling, the compressor and condenser fans start at the same time. However, the head pressure has fallen to a point that will cause the face dampers to be closed and the bypass dampers to be fully open. See Figure 13-9. As the compressor continues to operate, the head pressure rises, especially if the ambient air temperature is high enough to require more condenser cooling. When the head pressure has risen enough, the dampers gradually turn to allow more air to go through the condenser for proper cooling. If the ambient air temperature should fall and less condenser cooling is needed, the dampers modulate to close the face damper and open the bypass damper.

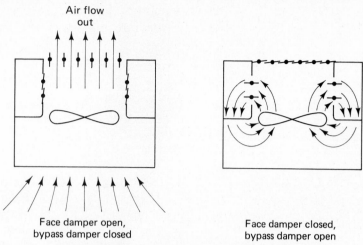

Air flow
out

Face damper open,
bypass damper closed

Face damper closed,
bypass damper open

**Figure 13-9.** Damper control operation.

## TEMPERATURE-ACTUATED MODULATING WATER VALVES

**Figure 13-10.** Temperature-actuated modulating water valve (Courtesy of Johnson Controls, Inc., Control Products Division).

*Temperature-actuated modulating water valves* are designed to control the flow of water through a water-cooled condenser to maintain the desired head pressure. See Figure 13-10. They are available in several different temperature ranges, so the correct one must be used. The bulb is mounted directly in the water stream and acts in response to the water temperature. These valves are designed to open on a rise in temperature of the water.

*Operation.* When the condenser has been idle for a period of time, the water temperature has fallen to a point where the valve seat is closed and no water is flowing through the condenser. When the indoor thermostat demands cooling, the compressor starts compressing the refrigerant. As the compressor continues to operate, the refrigerant temperature increases to the point that more water is required to maintain the desired head pressure. With an increase in refrigerant temperature, the water temperature also rises. The sensing bulb feels this increase in temperature and causes the valve needle to move off its seat and allow water to flow through the condenser. As the desired condensing pressure is reached, the valve begins to throttle, reducing the amount of water flowing through the condenser. The water valve modulates to maintain this head pressure until the unit is stopped again. Then the water valve slowly closes, stopping the flow of water through the condenser. When the condenser has cooled sufficiently, the valve closes off and completely stops the flow of all water.

## PRESSURE-ACTUATED WATER-REGULATING VALVE

*Pressure-actuated water regulating valves* are also modulating water-regulating valves. They are designed for use with a particular refrigerant so the correct one for the application must be used. See Figure 13-11. These valves act in direct response to the head pressure inside the system. They are connected to the high-pressure refrigerant line at the outlet of the receiver tank and open on a rise in refrigerant pressure.

**Figure 13-11.** Pressure-actuated water-regulating valve (Courtesy of Johnson Controls, Inc., Control Products Division).

*Operation.* When the system has been off for a period of time, the head pressure has fallen to a point that the valve seat is closed and no water is flowing through the condenser. When the indoor thermostat demands cooling, the compressor starts compressing the refrigerant, causing an increase in the pressure. As the head pressure increases the valve senses this increase and starts opening. When the correct amount of water is flowing through the condenser to maintain the desired head pressure, the valve begins to throttle and reduce the amount of water flowing through the condenser. The valve modulates to maintain this head pressure until the compressor stops running. Then the pressure starts to fall and the valve starts to close off, reducing the amount of water flowing. When the head pressure has fallen sufficiently, the valve closes off completely.

**THREE-WAY WATER-REGULATING VALVES**

*Three-way water-regulating valves* are designed for use with cooling towers. They are connected so that a part of the water is diverted back to the tower as the head pressure is decreased. See Figure 13-12. They are connected to the high-pressure line, usually at the outlet of the receiver tank. They act in direct response to the refrigerant pressure. They are designed for use with different types of refrigerant, so the correct valve for the application must be used.

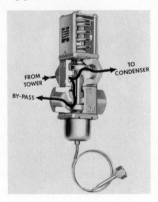

**Figure 13-12.** Three-way water regulating valve (Courtesy of Johnson Controls, Inc., Control Products Division).

*Operation.* When the compressor has been off for a while, the valve is open to the bypass port, allowing all the water to return to the tower. When the indoor thermostat demands cooling, the compressor starts, and the refrigerant pressure is increased to the point that the valve begins to modulate toward the open position at the condenser port and the close position at the bypass port. When the desired operating pressure is reached, the valve modulates allowing part of the water to go through the condenser and the rest to be diverted back to the tower. When the compressor stops, the valve modulates to open the bypass port and close the condenser port. When the pressure has fallen sufficiently, the valve completely closes off the condenser port and opens the bypass port.

## MOTOR-ACTUATED THREE-WAY WATER-REGULATING VALVE

*Motor-actuated three-way water-regulating valves* are designed to be used with a modulating motor, matching thermostat and the proper linkage. See Figure 13-13. They are piped into the system so that a part of the water may be diverted back to the cooling tower. The pressure control uses a potentiometer to match the one in the motor. It is connected to the outlet of the receiver tank to sense the pressure inside the high side of the system. They are used to maintain the desired head pressure by diverting some of the water back to the tower rather than delivering all of it through the condenser.

**Figure 13-13**. Motor-actuated three-way mixing valve (Courtesy of Johnson Controls, Inc., Control Products Division).

*Operation.* When the system is at rest, the head pressure is equal to the ambient temperature of the unit and the valve is fully open to the bypass port. When the compressor starts, the head pressure begins to rise. With this increase in head pressure, the pressure is sensed by the modulating pressure control. A signal is sent to the valve motor, which responds by opening the valve to the condenser port, admitting water to the condenser. As the unit continues to operate, the head pressure rises slowly and the valve opens, admitting more water to the condenser. When the balance point between the water flow and the head pressure is reached, the valve holds this position until a change in pressure is sensed by the pressure control. Then the valve modulates to allow the correct amount of water to flow through the condenser. When the compressor is stopped, the head pressure begins to fall, and the valve begins to modulate toward the open position for the bypass port.

## REVIEW QUESTIONS

1. When are head, or discharge, pressure controls necessary?
2. Define a head-pressure control.
3. What is the purpose of a cooling tower-fan control?
4. What control is used to control the fan speed on air-cooled condensing units?
5. Name two types of two-speed condenser-fan controls.
6. On a system using a two-speed condenser fan, what speed is the fan in when the unit is first started?
7. What device is used to cycle the condenser fan on and off to maintain the desired head pressure?
8. At what pressure does the low-ambient kit contacts open?
9. Why are both face and bypass dampers used on condenser-air-modulation systems?
10. What is the purpose of water regulating valves on water-cooled condensers?

# 14

# Modulating Motors and Step Controllers

The on-off type of control system is probably the least expensive type to use. When the thermostat senses a need for conditioned air, it turns on; when the need is satisfied, it turns off.

A significant improvement over the on-off control can be produced by dividing the conditioning load into a number of separate elements and keeping enough of them on continuously to provide an even flow of conditioned air to the area. This is done by using modulating motors and step controllers to turn on and off the number of conditioning elements required to maintain a constant temperature.

**DEFINITIONS** The following are the accepted definitions of a modulating motor and a step controller.

**Modulating Motor:** A *modulating motor* is a motor that will change position in response to a signal when the proper signal is received.

**Step Controller:** A *step controller* consists of a series of switches (steps) operated by a motor.

**MODULATING MOTOR** Modulating motors are available for a variety of different functions, such as modulating, spring-return; reversing two-position; reversing proportional; and two-position, spring-return motors.

**Figure 14-1.** Modulating spring-return (Courtesy of Johnson Controls, Inc., Control Products Division).

**Figure 14-2.** Wiring diagram for on-off or floating operation (Johnson Controls, Inc., Control Products Division).

**Modulating, Spring-Return:** A *modulating, spring-return motor* is generally used to operate dampers and valves. It is equipped with helical spring, which returns the motor shaft to the normally de-energized position. See Figure 14-1. Depending upon the system's requirements, the motor will go either to the full open or full closed position on a power interruption. It has a fixed 90° stroke. It will travel the full 90° stroke in 34 seconds. It is available for use on either 120-volt or 24-volt control systems and may be wired for either on/off or floating operation to increase its use. See Figure 14-2.

### Reversing Two-Position and Proportional

*Reversing two-position and proportional motors* are designed to operate valves and dampers. A selection of torque output, depending upon the timing required, permits correct selection of the motor for the job. See Figure 14-3. The controllers used with these motors may be two-position SPDT or modulating devices. These motors require 24-volt electrical power to operate in both directions. If the power is interrupted, the motor remains in that position until the power is restored. The dampers or valves operated by these motors may be either fully closed, fully open, or modulating, depending upon the requirements and the type of controller used.

**Figure 14-3.** Reversing two-position modulating motor (Courtesy of Johnson Controls, Inc., Control Products Division).

### Two-Position, Spring-Return Motors

*Two-position, spring-return motors* are used to operate dampers or valves where it is desirable to return the controlled device to the normally de-energized position on a power failure or interruption. It requires 60 seconds for this motor to make a full 180° cycle. These motors require 24-volt electrical power for operation. See Figure 14-4.

**Figure 14-4.** Two-position, spring-return modulating motor (Courtesy of Johnson Controls, Inc., Control Products Division).

*Operation.* The modulating motor circuit operates to position the controlled device (usually a damper motor or motorized valve) at any point between fully open and fully closed (except for two-position motors); this will proportion the delivery to the need indicated by the controller mechanism. See Figure 14-5.

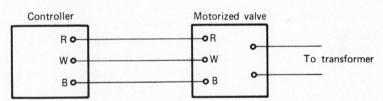

**Figure 14-5.** Modulating motor circuit-field wiring.

The power unit is a low-voltage capacitor motor, which turns the motor shaft by means of a gear train. Limit switches are used so that the rotation is limited to the rating of the motor. The gear train and all other moving parts are immersed in oil to eliminate the need for periodic lubrication and to ensure long, quiet service.

The power unit is started, stopped, and reversed by the single-pole–double-throw contacts of the balancing relay. See Figure 14-6. The balancing relay consists of two solenoid coils with parallel axes, into which are inserted the legs of a U-shaped armature. The armature is pivoted at the center so that it can be tilted by changing the magnetic flux of the two coils.

A contact arm is fastened to the armature so that it will touch one or the other of two stationary contacts as the armature moves back and forth on its pivot. When the relay is in balance, the contact arm floats between the two contacts, touching neither of them.

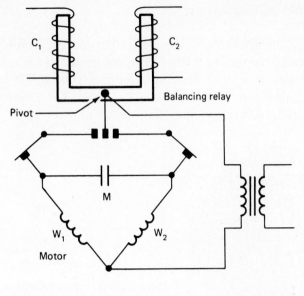

**Figure 14-6.** Diagram of balancing relay and motor circuit.

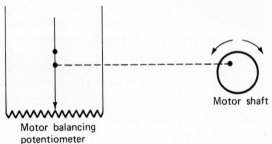

**Figure 14-7.** Schematic diagram of balancing potentiometer.

A balancing potentiometer is included in the motor. The potentiometer is electrically identical to the one in the controller. The finger is moved by the motor shaft so that it travels along a coil and establishes contact wherever it touches. See Figure 14-7.

Figure 14-6 illustrates how a balancing relay is made. As the relay is used in the modutrol circuit, the amount of current passing through the coils is governed by the relative positions of the controller potentiometer and the motor-balancing potentiometer. See Figure 14-8. Thus, when equal amounts of current are flowing through both coils of the balancing relay, the contact blade is in the center of the space between the two contacts and the motor is at rest. When the finger of the controller potentiometer is moved, more current flows through one coil than the other, and the relay is unbalanced. The relay armature is then rotated so that the blade touches one of the contacts and the motor runs in the corresponding direction.

The contact made by the balancing relay can only be broken if the amount of current flowing through $C_1$ is made equal to that amount flowing through $C_2$. This is brought about by the motor-balancing potentiometer linked to the motor shaft. As the motor rotates, it drives the finger of the motor-balancing potentiometer toward a position that equalizes the resistances in both legs of the circuit.

There is a definite position of the finger for each position of the motor shaft through its complete rated degrees of rotation. For example, when the motor shaft is rotated 40° from one of its extremes (25%

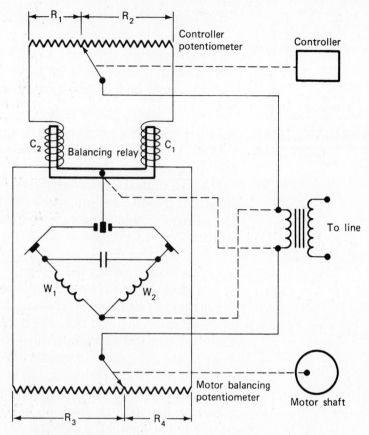

**Figure 14-8.** Diagram of control circuit in balanced condition.

**Figure 14-9.** Mounted auxiliary switches (Courtesy of Johnson Controls, Inc., Control Products Division).

of its arc), the finger is at $33\frac{3}{4}$ ohm, a value that lies 25% of the distance from the corresponding extreme of the coil resistance.

Figure 14-8 shows an instantaneous condition in which the current is flowing from the transformer, through the potentiometer finger, and down through both legs of the circuit. In the positions shown, the thermostat potentiometer finger and the motor-balancing potentiometer finger divide their respective coils so that $R_1 = R_4$ and $R_2 = R_3$. Therefore, $R_1 + R_3 = R_2 + R_4$ and the resistances on both sides of the circuit are equal. The coils $C_1$ and $C_2$ of the balancing relay are equally energized, and the armature of the balancing relay is balanced. The contact arc is floating between the two contacts, no current is going to the motor, and the motor is at rest.

Modulating motors may be used for the following applications:

1. *Air-flow diversion:* Where a parallel air flow pattern is used, a diverting damper is used to direct the air flow through either the heating unit or the cooling unit.
2. *Air-flow changeover:* Where a resistance damper is used to decrease the air flow on the heating cycle.
3. *Ventilation control:* Where provision is made for introducing outdoor air into the system during the cooling season but not during the heating season.

4. *Zoning:* Where both heating and cooling may be desired at the same time but in different areas.

5. *Valve operation:* Where steam or water may need to be directed in varying amounts, such as steam or hot-water heating coils or condenser-cooling water to a condensing unit.

**AUXILIARY (END) SWITCHES**

*Auxiliary switches* are designed to be mounted on the shaft of modulating motors. See Figure 14-9. The motor operates the switches at predetermined motor output shaft positions. Typical applications include capacity control on multistage refrigeration systems, multispeed fan controls, staging of multiple fuel-oil nozzles on boilers, and other such capacity-change requirements. These switches use mercury bulb contacts or micro switches for this purpose.

*Operation.*   When the system is off and the motor is positioned so that all of the switch contacts are open, nothing will be running. When the thermostat demands operation, the modutrol motor starts turning in the direction that will close the contacts. As the motor moves farther to the full demand position, more of the switches are closed to energize more of the equipment to handle the demand of the building or space being treated. When the motor has moved to the full demand position, all the switches are closed. As less equipment is needed, the motor starts turning in the opposite direction, and at each of the designated degrees of rotation of the shaft a switch is opened stopping a part of the equipment. This reduces the cost of operation while maintaining the design conditions inside the space. When the thermostat is satisfied, the motor returns to the off position and remains there until the thermostat again demands operation of the equipment.

**STEP CONTROLLERS**

*Step control* is appreciably better than on-off control because it meets load requirements continuously. See Figure 14-10. Temperature variations are substantially greater with on-off control under identical circumstances. However, the modulating effect of step control is strictly dependent on the number of steps provided. Ideally, a sufficient number of steps should be provided to hold the coil's air-temperature rise or fall to within 5°F (2.78°C) per step.

A step controller consists of a series of switches operated by an actuator. The actuator seeks a switch position corresponding to the controller requirements, thereby closing the correct number of switches. See Figure 14-11. The switches can directly energize the conditioning elements or control these elements through contactors.

The actuators on these units usually have a modulating type of motor and operate from a modulating thermostat. Some are of the proportioning type and are controlled by a staged (step) controller. Many types have a reversible motor-driven cam and step-switch assembly with limit switches, feedback potentiometer, recycle relay, multiple tapped transformer, and terminal strips for connecting the control circuit. Some models have up to ten adjustable switches and they may be mounted in any position.

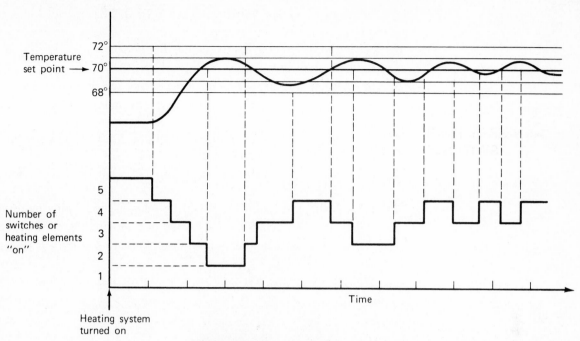

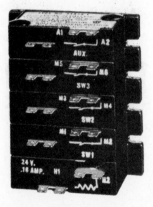

**Figure 14-10.** Temperature variation with step control.

**Figure 14-11.** Step controller (Courtesy of White-Rodgers Division, Emerson Electric Co.).

Step controllers may be used for control of any electric-heating unit containing elements that can be divided into separate circuits. Also, they may be used to start or stop refrigeration compressors that are connected in tandem to produce the necessary capacity. Compressor unloaders are also controlled with this control.

*Operation.* When the thermostat is satisfied, the step controller motor is positioned so that all the switch contacts are open and no equipment is running. When the thermostat demands operation, the step-controller motor begins to move to close the switch contacts in sequence. As the motor rotates farther, more of the switches are closed until all of them are demanding operation of the equipment. As the thermostat becomes more satisfied, the motor begins to rotate in the opposite direction and opens the switches as the motor reaches a certain degree of rotation. When the thermostat is completely satisfied, the step-controller motor returns to the at-rest position and remains there until another demand is made on the system.

## REVIEW QUESTIONS

1. Name the different types of modulating motors.
2. Define a step controller.
3. In what applications are modulating, spring-return motors used?
4. What device is used to return spring-return motors to the de-energized position?
5. What is required for operation in both directions of the reversing two-position and proportional motors?
6. On two-position, spring-return motors, of what does the power unit consist?
7. What is the purpose of the balancing relay in modulating motors?
8. Where are the potentiometers located in a modulating control system?
9. When is the balancing relay positioned so that no current is flowing to the motor windings?
10. Which provides the best comfort, step control or on-off control systems?

# 15

# Timers and Time Clocks

Timers and time clocks are devices that are used for the automatic and economical operation of air conditioning, heating, and refrigeration equipment. They are designed with a variety of contact arrangements and contact current ratings. They may be set to provide almost any on-off cycle arrangement desired. If the time control you need is not immediately available, most manufacturers will welcome the opportunity to satisfy your time control requirements specially.

**DEFINITIONS** The following are definitions of timers and time clocks:

**Time Clock:** The *time clock* is an electrically operated control used to govern the various cycling functions of equipment over a given period of time. See Figure 15-1.

**Figure 15-1.** Time clock (Courtesy of Intermatic Incorporated).

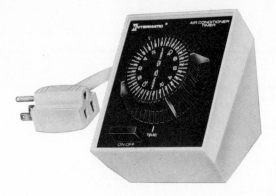

**Figure 15-2.** Timer (Courtesy of Intermatic Incorporated).

**Timer:** The timer is similar to the time clock except that its functions are limited to less complex systems. Generally, they control the on-off cycles of only one appliance, such as window air conditioning units, for operation with one or two on-off cycles per 24 hours. They generally have a receptacle for plugging in the controlled appliance. See Figure 15-2.

*Operation.* To help in understanding how these controls operate you should first become acquainted with the interior components of a time clock. The gears and parts are heavy and rugged enough to withstand prolonged normal usage. See Figure 15-3. They are designed and built to function properly even under adverse conditions.

The high-powered switching mechanism is made of channeled brass U-beam blades, which are operated by a rugged cam to give instant and positive make and break action. See Figure 15-4. The contacts are usually self-cleaning and made of a special alloy to prevent pitting. They are rated to carry an inrush of current ten times their normal amperage rating without arcing or sticking.

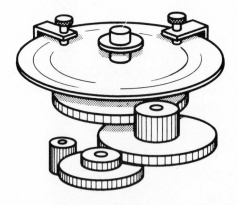

**Figure 15-3.** Gears.

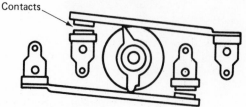

**Figure 15-4.** Switching mechanism.

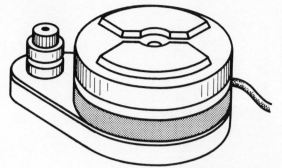

**Figure 15-5.** Industrial motor.

Time clocks are usually powered by heavy duty industrial-type motors. See Figure 15-5. These are synchronous-type motors, extremely quiet in operation and self-lubricating. They never need service or attention and are practically immune to adverse temperature and humidity conditions.

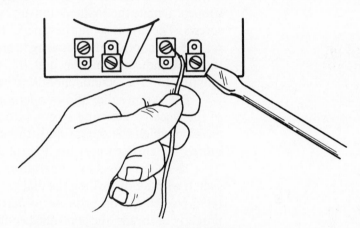

**Figure 15-6.** Terminal board.

The terminal board is designed for fast and easy wiring; there is plenty of room for hands and wiring procedures. See Figure 15-6. The control dial is divided accurately into units of time. See Figure 15-7. Most timer control dials are painted with contrasting colors to aid in the placement of the trippers. The trippers are adjustable to provide the desired

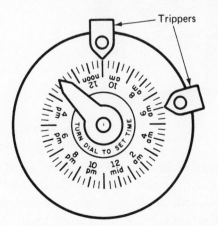

**Figure 15-7.** Control dial.

**Figure 15-8.** Time clock showing skipper dial (Courtesy of Intermatic Incorporated).

on and off periods. These dials are designed to cover 24 hours. The trippers actuate the contacts and are fastened to the rotating dial by thumb screws. The dial is set at the correct time by pushing the dial toward the rear of the control and rotating it until the correct time corresponds to the time indicator on the dial. It is necessary, however, to read the manufacturer's recommendations before attempting to set the dial.

The skipper dial allows operations to be skipped on selected days of the week by inserting screws into the proper holes in the dial for the days on which the operation is to be omitted. Seven-day clocks are required for this feature. See Figure 15-8.

The hand trip allows operation of the equipment through the timed switch without disturbing the dial settings. See Figure 15-9. The trip may be moved to the on position to determine if the system is functioning properly or to start the controlled equipment when an extra load or similar demand is placed on the equipment.

**Figure 15-9.** Hand trip (Courtesy of Intermatic Incorporated).

In operation the time-clock motor is wired in parallel with the switching contacts. As the motor rotates the dial and trippers, the contacts are opened or closed, depending upon the desired operation at that time. A simple diagram is shown in Figure 15-10. Most time clocks are provided with a spring-wound carry-over mechanism. The carry-over

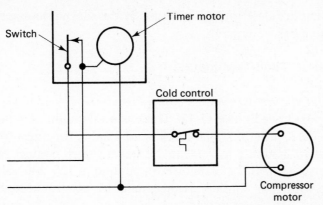

**Figure 15-10.** Simple wiring diagram of defrost circuit.

mechanism keeps the time clock on schedule for 36 hours during power failures. When the power resumes, the carry-over automatically rewinds itself, providing ideal control in areas where power breakdown frequently occurs.

A time clock provided with all of the devices mentioned previously can be used to control almost any operation desired. The following are some of the more common uses of time clocks:

1. Heating control
2. Ventilation control
3. Air conditioning control
4. Defrost control

Timers provide automatic control over a great variety of functions such as cycling of window air conditioning units, heating units, fans, lights, and anything requiring periodic on-and-off cycles. Through the use of relays, timers may be used to provide the same basic functions as the basic time clock. The 24-hour dial is divided into half-hour increments with permanently attached on and off trippers, permitting the desired on-and-off periods in each 24 hours. See Figure 15-11.

The following are some suggested uses for the timer:

1. Time-delay relays
2. Motor control (to 1 horsepower)
3. Pump control
4. Heating control
5. Domestic refrigeration control

**Figure 15-11.** Timer dial (Courtesy of Intermatic Incorporated).

**DEFROST TIME CLOCKS**

The following are descriptions of some of the more popular types of defrost time clocks.

### Time-Initiated–Time-Terminated:

A *time-initiated–time-terminated control* can be set for up to six defrost periods per day. See Figure 15-12. There is a minimum of 4 hours between each defrost period. The clock is adjustable for times from 4 to 110 minutes in 2-minute increments. It has a switch contact rating of 30 amps per pole, with 2-horsepower maximum at 120–240 volts ac. It is available with four contact arrangements for electric heat, hot gas, or compressor shutdown defrost.

**Figure 15-12.** Time-initiated–time-terminated defrost timer (Courtesy of Precision Multiple Controls).

*Operation.* The time and the termination of the defrost cycles for units on which these controls are used are both set on the dial of the timer with timer pins. See Figure 15-13. The defrost cycle is started at

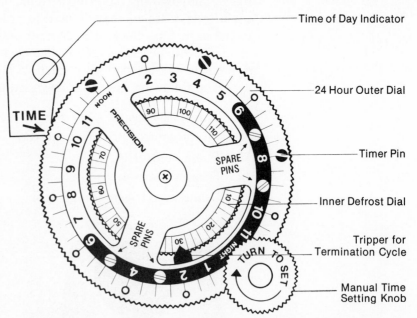

**Figure 15-13.** Timer dial (Courtesy of Precision Multiple Controls).

a given time or times each day and it is terminated at the same given time or times each day. This control does not take into consideration whether or not the frost is completely removed or if the unit stays in defrost longer than necessary.

### Time-Initiated–Temperature-Terminated

The defrost periods for which the *time-initiated–temperature-terminated control* is used are adjustable from one to six per day with a minimum of 4 hours between each defrost cycle. See Figure 15-14. The defrost periods are adjustable for times from 4 to 110 minutes each with 2-minute increments. The control has a switch contact rating of 30 amps per pole with a 2-horsepower maximum, using 120–240 volts ac. The control has three contact arrangements for electric heat, hot gas, or compressor shutdown defrost cycles.

This control is designed for use with an external temperature or pressure sensor. There is an inner dial, which provides a backup defrost-termination feature and protects the system against a malfunction.

**Figure 15-14.** Time-initiated–temperature-terminated defrost timer (Courtesy of Precision Multiple Controls).

*Operation.* At a given, preselected time, each defrost period is started by the timer. When the temperature of the sensor on the timer reaches a predetermined temperature, the control terminates the defrost cycle. Thus, the length of the defrost cycle varies according to the amount of frost on the coil. Should something happen to the temperature sensor, a backup timer function takes the system out of defrost after a preselected period of time has lapsed.

### Time-Initiated–Pressure-Terminated

The *time-initiated–pressure-terminated control* provides for one to four defrost cycles per day or one to six in 24 hours. See Figure 15-15. The length of these defrost cycles is adjustable from 4 to 110 minutes in 2-minute increments with a minimum of 4 hours between successive defrost cycles. The switch contact ratings are for 30 amps per pole, 2 horsepower maximum at 120–240 volts ac. Three contact arrangements are available for hot gas, electric heat, or compressor shutdown for defrost. The control is supplied with an adjustable-pressure cut-in set-

**Figure 15-15.** Time-initiated–pressure-terminated defrost timer (Courtesy of Precision Multiple Controls).

ting dial, which is calibrated from 35 to 110 pounds per square inch (344.22 to 860.97 kilopascals). It is also provided with a timed mechanical backup termination feature, which protects the system against sensor malfunctions.

*Operation.* The defrost cycle is initiated each time by the time clock. However, it takes a predetermined pressure to terminate the defrost period. This pressure is exerted by the refrigerant inside the system in response to the temperature of the evaporating coil being defrosted. When this pressure has risen to the cut-out setting of the defrost control, the system will start in the normal cooling cycle. Should this pressure sensor fail, the timed backup mechanical feature prevents damage to the equipment by switching the system to the normal cooling cycle.

### Multicircuit Defrost Controls

A *multicircuit defrost control* is a multicircuit timing device that will operate from 1 to 24 switches individually. See Figure 15-16. These are positive snap-acting switches that provide positive and immediate switching of the circuits. These times were designed to control a series of operations with a variable time sequence, such as a refrigeration defrost cycle. They are field adjustable timers that combine ease of programming with accuracy and reliability.

With this control each switch is independently controlled by the combined action of two separate time shafts. This arrangement allows for extremely short on-cycles with extremely long off-cycles, such as a 2-minute on-cycle once each day. It also permits the switches to be cycled independently at selected times for independently adjusted on-cycles. For example, one switch can be adjusted to operate once each day with a 2-hour on-cycle and another switch can be set to operate eight times per day with a 10-minute on-cycle.

Each switch is actuated by an individually adjustable timer. This consists of a cam assembly having a drive gear arranged to rotate one revolution at a relatively high speed and then stop, giving one cycle to the switch. The starting of each cycle timer is determined by an individual

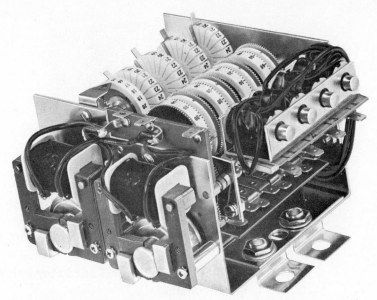

**Figure 15-16.** Multicircuit defrost control (Courtesy of Precision Multiple Controls).

timer rotating at the relatively slow speed of one revolution per 24 hours. Each program timer includes a molded dial with slots into which trippers can be inserted.

*Operation.* In operation, a tripper of the program timer serves to start the cycle timer, which then goes through one cycle and comes to rest. Adjustment of the timer determines the length of the on-cycle and insertion of the trippers in the program timer determines when the on-cycles occur.

### 24-Hour Time-Switch Skip-a-Day Series:

The *24-hour time-switch skip-a-day* clock is designed to allow the operator to omit operation on Saturday, Sunday, or any other selected days. See Figure 15-17. The timer is equipped with a 24-hour dial with the skip-a-day feature included. It is equipped with either SPDT or DPST switching arrangements. The contacts are rated 30 amps tungsten with 2-horsepower maximum using 120–240 volts ac. This timer is equipped with a manual on-off switch, which allows the system to be hand operated without disturbing the scheduled settings of the clock.

*Operation.* System operation is activated by setting the trippers for each day that operation is desired. However, when the day that has been selected for the system not to operate comes around, the tripper will not start the system as usual. When this 24-hour period has passed, the system will again function as desired according to the placement of the trippers.

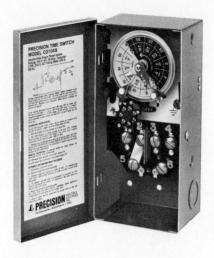

**Figure 15-17.** Twenty-four-hour 24-volt time switch for simple thermostat setback with skip-a-day (Courtesy of Precision Multiple Controls).

### 7-Day Time-Switch Four-Pole with Carryover Mechanism:

This *7-day time-switch* control permits 7-day operation with a different program for each operating day using a spring-wound carryover mechanism. See Figure 15-18. The purpose of the spring-wound carryover feature is to keep the control on schedule during a power failure. It continues to run up to a maximum of 10 hours on reserve power. When the power is restored, the carryover mechanism rewinds itself.

**Figure 15-18.** Seven-day time switch, four pole with carryover mechanism (Courtesy of Precision Multiple Controls).

*Operation.* This timer operates just the same as any standard timer. The only exception is that with a power failure, the clock maintains the set points at the desired times.

**WIRING DIAGRAMS**      The wiring diagrams in figures 15-19 through 15-24 are only suggestions for wiring time clocks into the circuit for different system requirements.

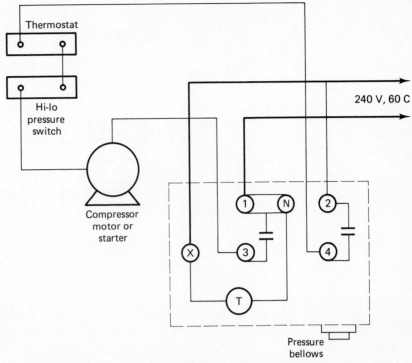

**Figure 15-19.** Wiring using 240-volt single-phase compressor motor. Breaking both sides of 240-volt line. (Courtesy of Paragon Electric Company, Inc.).

**Figure 15-20.** Wiring for compressor motor with magnetic starter. Clock motor independent of load circuit (Courtesy of Paragon Electric Company, Inc.).

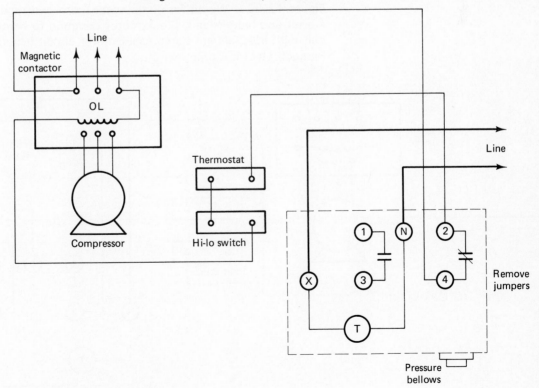

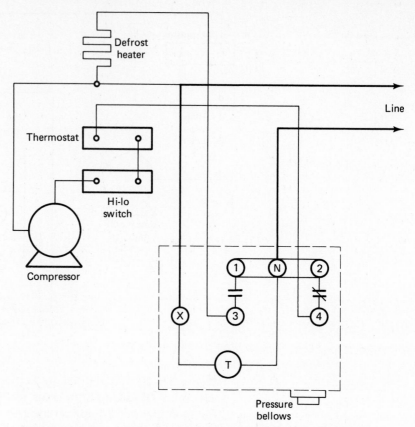

**Figure 15-21.** Wiring using single-phase line for electric heat system without magnetic starter (Courtesy of Paragon Electric Company, Inc.).

**Figure 15-22.** Wiring for compressor motor with magnetic starter and heater load. Clock motor common to heater circuit with independent compressor motor circuit (Courtesy of Paragon Electric Company, Inc.).

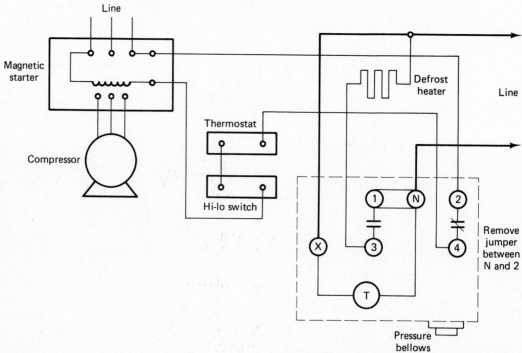

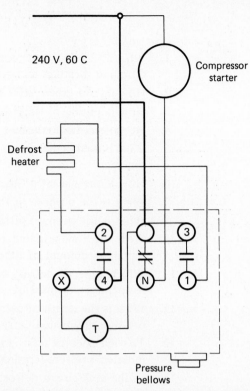

**Figure 15-23.** Wiring using 240-volt single-phase line for electric heat system with split-load defrost heaters (Courtesy of Paragon Electric Company, Inc.).

**Figure 15-24.** Wiring using 240-volt single-phase line for electric heat system with split-load defrost heaters (Courtesy of Paragon Electric Company, Inc.).

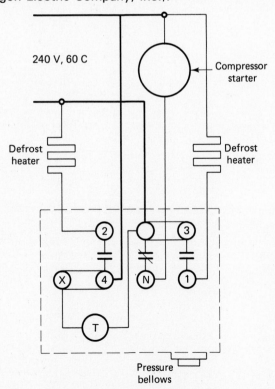

## REVIEW QUESTIONS

1. What devices are used for automatic and economical operation of air conditioning, heating, and refrigeration equipment?
2. How does a timer differ from a time clock?
3. How is pitting of timer switch contacts prevented?
4. How often should the motors used in timers and time clocks be lubricated?
5. What is used in timers and time clocks to switch the contacts at the desired time?
6. How is the dial of a timer or time clock set at the correct time?
7. What is the purpose of the skipper dial on timers and time clocks?
8. What is the purpose of the hand trip on timers and time clocks?
9. With the time-initiated–time-terminated defrost control, what is the minimum amount of time between each defrost period?
10. Which type of defrost control is designed for use with an external temperature or pressure sensor?
11. What is the length of defrost cycles on systems using the time-initiated–pressure-terminated defrost control?
12. To what does the refrigerant pressure respond in a system using the time-initiated–pressure-terminated defrost control?
13. In the multicircuit defrost control, what controls each switch?
14. What determines the starting of each cycle timer in the multicircuit defrost control?
15. What does the 7-day time-switch four-pole with carryover mechanism permit?

# 16

# Defrost Systems

Defrost systems are used on two different types of systems. When a refrigeration unit producing below-freezing temperatures is operated for a period of time, a layer of frost or ice is built up on the evaporator. This layer of ice acts as an insulator, which decreases the efficiency of the equipment. When a heat-pump system is operating in the heating mode, frost or ice accumulates on the outdoor coil, which also reduces the efficiency of the heat pump unit and must be removed for satisfactory and economical operation. To employ a person to keep watch and defrost the coils would be too expensive; thus automatic defrost systems are used. The purpose is the same but the method and controls are somewhat different.

**DEFINITION** *Defrost systems* are a variety of controls used to accomplish the automatic defrosting of evaporating coils.

**OPERATION** The simplest defrost system uses a low-pressure control to cycle the compressor. The low-pressure control is set low enough to maintain the lowest temperature required in the refrigerated space. This device has primary control over the compressor. Operation of the compressor is started only after the temperature and the corresponding refrigerant pressure have risen sufficiently to cause some of the frost to melt during each off-cycle of the equipment.

This type of defrost system is not very efficient because not all the ice is melted during each off-cycle of the equipment. When the low-pressure control is used for this purpose, the complete system must be shut down periodically to defrost the system manually. This system is not the most economical; it is also inconvenient.

The most economical type is one that automatically and completely defrosts the evaporator during each defrost cycle of the equipment. This type of system is usually time-initiated and temperature-terminated. It allows a more even temperature during the busy time of the day, and the defrost period is accomplished during a period of less activity.

**CLASSIFICATION OF DEFROST SYSTEMS**

There are two main classifications of defrost systems: (1) the hot-gas defrost and (2) the electric defrost. The hot-gas defrost system is discussed first.

### Hot-Gas Defrost System

The hot-gas defrost is the most rapid and economical method of automatic, positive defrosting. Refrigerant vapor, superheated by compression, is circulated through the evaporating coil to provide a continuous supply of defrosting heat.

During the defrost cycle, the evaporator and condenser fans are turned off. Some means of reversing the flow of refrigerant is used to direct the hot gas into the evaporating coil to melt the frost. The reversing is usually done with a solenoid valve or a four-way reversing valve, such as those used on heat pump systems.

*Types of Hot-Gas Defrost Systems.* The different types of defrost systems are designated by the types of controls used to cause the defrost cycle, such as time-initiated–time-terminated, time-initiated–temperature-terminated, time-initiated–pressure-terminated, and air-pressure differential. The following is a discussion of these different types of control systems.

*Time-Initiated–Time-Terminated:* The *time-initiated–time-terminated* system uses a time clock designed especially for this type of use. These time clocks were discussed in Chapter 15. The trippers on the time clock are set to provide a defrost cycle at a given time, or times, each day, depending upon the requirements of the system. The time clock initiates a defrost at the desired time, and the defrost cycle continues until the time clock terminates the cycle. See Figure 16-1. The defrost intervals should be determined by box usage so that defrosting is completed and the evaporating coil is free of frost just prior to heavy usage.

*Time-Initiated–Temperature-Terminated:* The *time-initiated–temperature-terminated* defrost system uses a time clock designed for this purpose. These time clocks were discussed in Chapter 15. The on trippers are set to start the defrost cycle or cycles at a predetermined time each day. The system remains in the defrost cycle until the temperature and

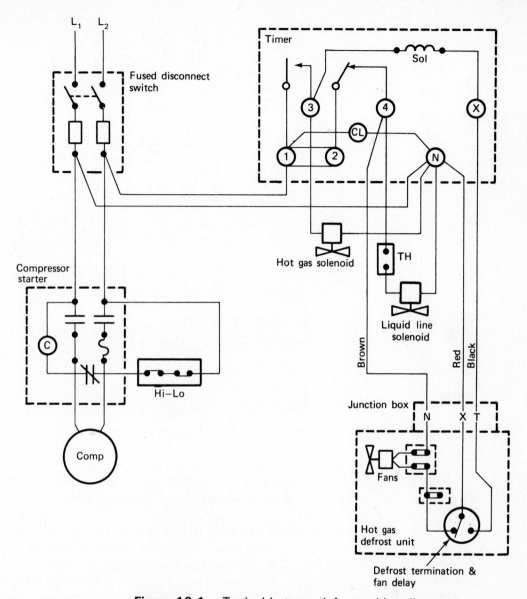

**Figure 16-1.** Typical hot-gas defrost wiring diagram.

corresponding pressure of the evaporator have increased to the point that indicates that all the frost has melted from the evaporating coil. This temperature is sensed by a defrost termination control. At this point the defrost cycle is terminated and the system returns to the normal operating sequence. See Figure 16-2. The defrost intervals should be determined by box usage so that defrosting is completed and the evaporating coil is free of frost just prior to heavy usage.

*Time-Initiated-Pressure-Terminated:*   The *time-initiated–pressure-terminated* defrost system uses a time clock designed for this purpose. These time clocks were discussed in Chapter 15. The on trippers start the defrost cycle or cycles at a predetermined time each day. The system remains in defrost until the refrigerant pressure inside the system has

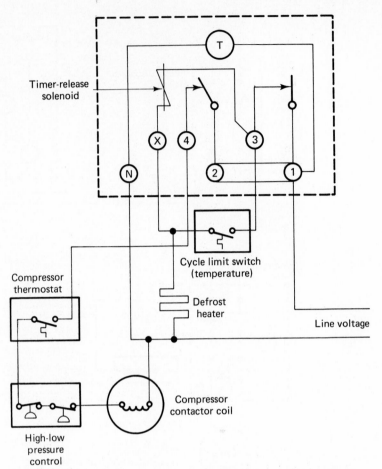

**Figure 16-2.**  Typical wiring diagram for a time-initiated–
temperature-terminated defrost system.

increased to the point that indicates that all the frost has been melted
from the evaporating coil. At this point the defrost cycle is terminated
by the pressure control and the system returns to the normal operating
sequence. See Figure 16-3. The defrost intervals should be determined
so that defrosting is completed and the evaporating coil is free of frost
just prior to periods of heavy usage.

### Heat-Pump Defrost Controls

When heat-pump systems are operating in the heating mode, the outdoor coil becomes the evaporator and refrigerant is evaporating in it.
When the temperature of the outdoor coil drops below 32 °F (0 °C), frost
and ice begin to form on the coil. Ice and frost reduce the efficiency
of the unit and must be removed. This is the purpose of the defrost system
on heat pump units.

Frost is removed by the controls that cause the reversing valve to
change positions and direct the hot gaseous refrigerant into the outdoor
coil. When this shift takes place, both the indoor and outdoor fans are
stopped to aid in a rapid defrost of the unit.

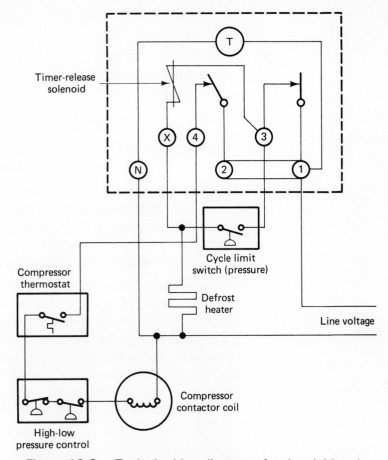

**Figure 16-3.** Typical wiring diagram of a time-initiated–pressure-terminated defrost system.

The most popular methods of automatic defrost initiation and defrost termination are (1) air-pressure differential across the outdoor coil, (2) outdoor coil temperature, (3) time, (4) time and temperature, and (5) solid-state defrost control.

*Air-Pressure-Differential (Demand) Defrost Systems.* *Air-pressure-differential defrost systems* use a diaphragm-type control to measure the air-pressure differential across the outdoor coil. They provide a defrost cycle on demand rather than by using a time clock. See Figure 16-4. These

**Figure 16-4.** Air-pressure differential defrost control (Courtesy of Ranco Controls Division).

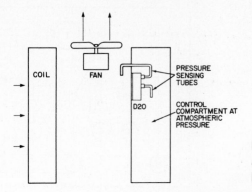

**Figure 16-5.** Typical installation of air pressure sensor (Courtesy of Ranco Controls Division).

devices use a slow make or break switch, which is actuated by a slack-bag diaphragm. Low-pressure taps permit operation on positive or negative pressure applications. At the calibration pressure, the switch closes, providing current to the defrost control. See Figure 16-5.

*Operation.*    As frost forms on the outdoor coil of the heat pump, the air-pressure differential across the outdoor coil increases. The control measures this pressure drop. As the pressure differential increases to a predetermined difference, the inflated slack-bag diaphragm operates a pilot-duty switch, which directs the initiation signal to the defrost control. The defrost control then stops operation of the indoor and outdoor fans, switches the reversing valve, and makes provision for the indoor resistance heaters to operate when needed. See Figure 16-6.

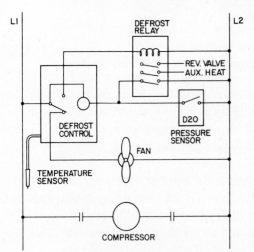

**Figure 16-6.** Typical heat pump installation (Courtesy of Ranco Controls Division).

*Outdoor-Coil Temperature.*    In the *outdoor-coil temperature* system the defrost cycle is automatically initiated by a defrost thermostat. See Figure 16-7. This is a double-bulb thermostat that senses a reduction in the efficiency of the unit and initiates the defrost cycle. Time, outdoor air temperature, wind, and atmospheric conditions have no affect on this control.

The signal that causes initiation of the defrost cycle is caused by an increase in the temperature difference between the ambient air temperature around the outdoor coil and the temperature of the coil itself with ice on it, as compared to the temperature difference of an ice-free coil at the same outdoor temperature.

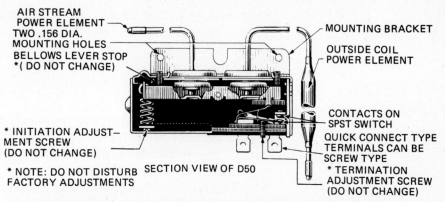

AIR STREAM
POWER ELEMENT
TWO .156 DIA.
MOUNTING HOLES
BELLOWS LEVER STOP
*( DO NOT CHANGE)

MOUNTING BRACKET

OUTSIDE COIL
POWER ELEMENT

CONTACTS ON
SPST SWITCH

QUICK CONNECT TYPE
TERMINALS CAN BE
SCREW TYPE

* TERMINATION
ADJUSTMENT SCREW
(DO NOT CHANGE)

* INITIATION ADJUST—
MENT SCREW
(DO NOT CHANGE)

SECTION VIEW OF D50

* NOTE: DO NOT DISTURB
FACTORY ADJUSTMENTS

**Figure 16-7.** Defrost thermostat (Courtesy of Ranco Controls Division).

*Operation.*    As the unit operates and the ice accumulation reaches the point that the temperature difference is great enough, the defrost cycle is initiated, the outdoor fan stops, the reversing valve is caused to change positions, and the auxiliary heating elements are ready to be energized. See Figure 16-8. As the compressor continues to run, the outdoor coil is warmed up and the ice is melted from the coil. The outdoor coil remains at about 32 °F (0 °C) until all the frost has been melted. The unit continues to operate in this mode until the temperature of the out-

**Figure 16-8.** Schematic wiring diagram of thermostatic defrost control (Courtesy of Ranco Controls Division).

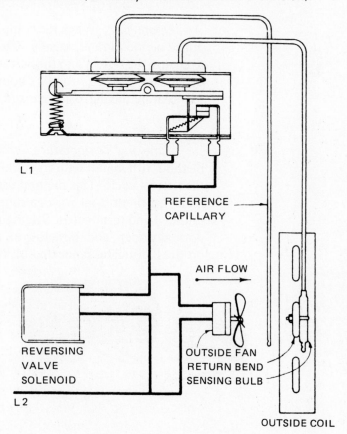

L 1

REFERENCE
CAPILLARY

AIR FLOW

REVERSING
VALVE
SOLENOID

OUTSIDE FAN
RETURN BEND
SENSING BULB

L 2

OUTSIDE COIL

door coil has reached the temperature needed to cause the reversing valve to change position. This temperature is usually about 60 °F (15.56 °C) on the preset control. At this temperature the defrost cycle is terminated. When the termination temperature has been reached, the system returns to the heating cycle automatically.

*Time.* A *time* defrost control usually initiates a defrost cycle after a given amount of compressor running time has elapsed. The defrost control uses a timer motor using contacts to provide the necessary switching action. See Figure 16-9.

This type of defrost system has several disadvantages. The unit may be switched into a defrost cycle when it is not needed, it may wait too long before initiating a defrost cycle, it may keep the unit in defrost too long, or it may not keep it in defrost long enough. These conditions will usually cause the unit to operate at less than its peak efficiency.

**Figure 16-9.** Defrost timer (Courtesy of Lennox Industries, Inc.).

*Operation.* When the compressor has been running for a predetermined period of time, usually 90 minutes, the defrost timer contacts close to change the system into the defrost mode. After a predetermined amount of time has elapsed, usually about 15 minutes, the timer contacts open, terminating the defrost cycle and placing the unit back into the heating mode.

*Time and Temperature.* *Time and temperature* is a very popular method with manufacturers for automatically initiating and terminating the defrost cycle. This method uses a combination of a clock timer and a defrost thermostat for determining when the defrost cycle should be initiated and terminated. See Figure 16-10. The thermostat contacts are normally open, and they close on a fall in temperature. The sensing bulb on the thermostat is located on one of the outdoor-coil tubes near the

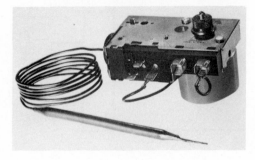

**Figure 16-10.** Defrost timer and thermostat (Courtesy of Ranco Controls Division).

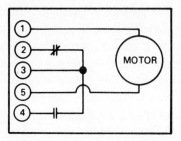

**Figure 16-11.** Defrost time-clock timer schematic (Courtesy of Lennox Industries, Inc.).

refrigerant outlet. When the sensing element determines that the outdoor-coil temperature has dropped to 32 °F (0 °C), the thermostat contacts close. These contacts are in electrical series with the NO contacts in the clock timer. See Figure 16-11. The clock motor is electrically parallel to the outdoor fan motor and runs only when the compressor is running. See Figure 16-12.

**Figure 16-12.** Wiring diagram of heat pump (Courtesy of Lennox Industries, Inc.).

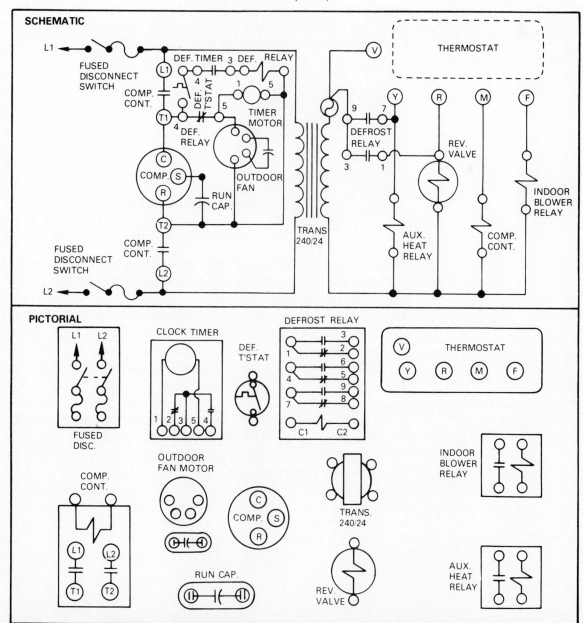

*Operation.* When a predetermined period of time has elapsed, usually 90 minutes (these timers are adjustable for either 30-, 60-, or 90-minute cycles), the timer motor closes its NO set of contacts. These contacts remain closed for only a few seconds during each cycle of the timing mechanism. If the defrost thermostat contacts close at the same time, the defrost relay coil is energized, which initiates a defrost cycle. If the thermostat contacts do not close at this time, indicating that a defrost cycle is not needed, the clock timer will start another cycle.

When the unit goes into a defrost cycle, the clock-timer motor and the outdoor fan are de-energized. The thermostat will keep the unit in the defrost cycle until the outdoor coil temperature has increased to approximately 65 °F (18.33 °C). At 65 °F the thermostat contacts will open, de-energizing the defrost relay and terminating the defrost cycle.

*Solid-State Defrost Control.* The *solid-state defrost control* method makes use of a solid-state control and two thermistors. See Figure 16-13. A thermistor is a device that changes in electrical resistance with a change in temperature. One thermistor is located near the outlet of the outdoor coil to sense the coil temperature at this point. The other thermistor senses the outdoor air temperature as it enters the coil.

**Figure 16-13.** Solid-state defrost control (Courtesy of Lennox Industries, Inc.).

*Operation.* As the outdoor air temperature drops below 45 °F (7.22 °C), frost will probably start forming on the outdoor coil. Frost causes an increase in the temperature difference between the two thermistors. When the temperature difference is 15 °F to 25 °F (8.4 °C to 14 °C), the defrost cycle is initiated.

In this particular type of control system, the coil of the outdoor fan and defrost relay is a 24-volt dc coil. It is energized through terminals R and R of the solid-state defrost control. See Figure 16-14. In this type of system a high-pressure control is used to terminate the defrost cycle. When the refrigerant pressure inside the outdoor coil reaches 275 pounds per square inch gauge (1990.24 kilopascals), the defrost cycle is terminated. See Figure 16-15. This amount of pressure is an indication that the temperature (approximately 124 °F or 51.11 °C) of the outdoor coil is high enough to melt all the frost that has formed on its surface. This control is located in the refrigerant liquid line in the outdoor unit.

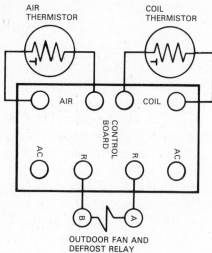

**Figure 16-14.** Solid-state defrost-control location (Courtesy of Lennox Industries, Inc.).

**Figure 16-15.** Defrost termination switch (Courtesy of Robertshaw Controls Co., Uni-Line Division).

*De-Ice Sensor Control.* The *de-ice sensor control* uses a pressure differential and a defrost-termination thermostat to initiate and terminate the defrost cycle. See Figure 16-16. The diaphragm sensors are located so that they sense the pressure drop across the outdoor coil. The thermostatic portion of the control is located on one of the outdoor-coil return bends. The complete process of defrost initiation and termination is fully automatic and is accomplished by use of this one control.

*Operation.* The pressure differential monitors the frost buildup by sensing the air-pressure change across the outdoor coil. The pressure differential initiates the defrost cycle, stopping the outdoor fan and changing the position of the reversing valve. The defrost cycle is terminated by the thermostat that senses the temperature of the outdoor coil. At this time the system is put back into the normal heating cycle by the control.

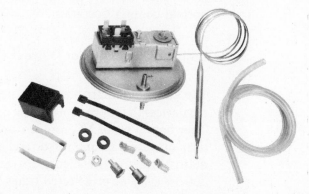

**Figure 16-16.** De-ice sensor control (Courtesy of Robertshaw Controls Co., Uni-Line Division).

### Electric Defrost System

When the *electric defrost system* is used, the compressor is shut down during the defrost period. An electric heater is energized to heat the evaporator and melt the ice. At the same time the evaporator and condenser fans are stopped. As the coil temperature is raised to a predetermined point, the defrost cycle is terminated, and refrigeration resumes in the usual manner. See Figure 16-17 for wiring diagrams.

A good-quality time control should be used to initiate the defrost cycle in both the hot-gas and the electric defrost systems.

The drain line should be heated in the refrigerated space to prevent freezing during the defrost period. This may be done with either an electric tape around the pipe or with the liquid line in contact with the drain line. The drain should be trapped outside the refrigerated space.

The liquid-line solenoid should be of good enough quality to prevent liquid refrigerant from entering the low side during the defrost cycle regardless of the thermal expansion valve setting.

Typical examples used for electric defrost systems are as follows:

1. Frozen food cabinets
2. Dairy cases
3. Vegetable cases
4. Beverage coolers
5. Ice cream display cases
6. Walk-in coolers

*Operation.*   These types of defrost systems use one of the time clocks discussed earlier in this chapter. When the defrost control senses that the evaporator needs to be defrosted, the control stops the compressor, evaporator fan, and condenser fan. It also energizes the electric resistance heater to defrost the evaporator coil. When the control senses that the frost has all been removed, the system is returned to its normal cooling operation.

### Domestic Refrigerator Defrost Controls

The automatic defrost controls used on domestic refrigerators are timing devices that operate when the unit is plugged into the electrical outlet. They have a set of NC contacts and a set of NO contacts. The NC contacts operate the compressor and all fans. The NO contacts control the defrost heaters. See Figure 16-18. These controls are set to provide a defrost cycle at preselected times during a 24-hour period. During the defrost period the compressor and all the fans are stopped and an electric resistance heater is energized. In some units a condensate drain heater is also energized to prevent condensation from freezing in the drain line in the freezer compartment.

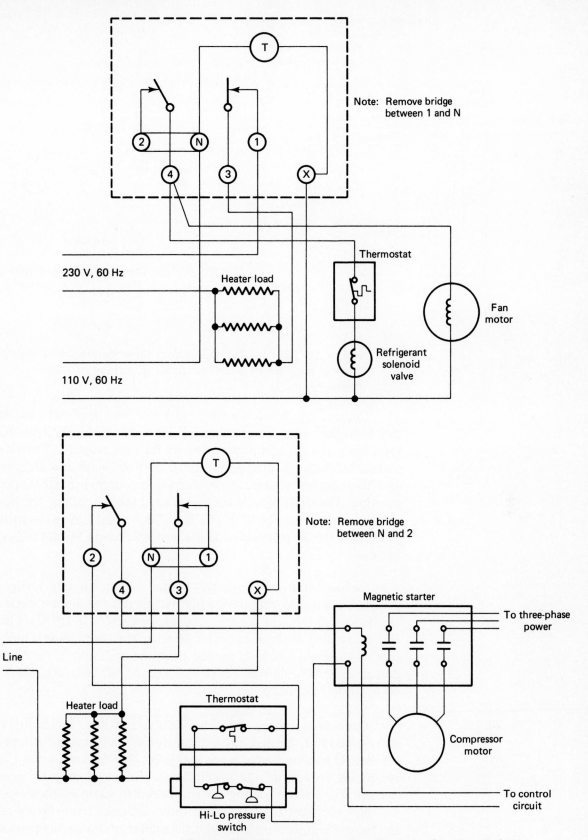

**Figure 16-17.** Typical wiring diagrams for electric heat defrost system.

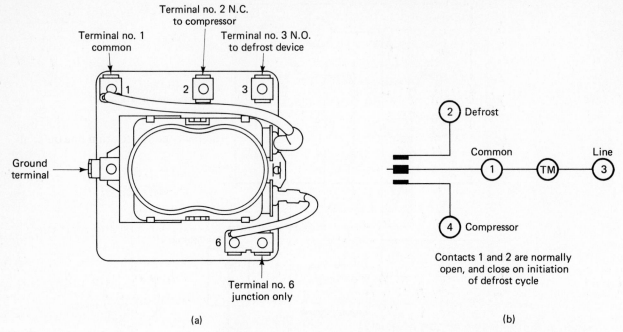

**Figure 16-18.** (a) Typical defrost timer connections and (b) internal diagram of a defrost timer (Courtesy of Gem Products, Inc.).

*Operation.* When the time clock reaches the preset time for the unit to be defrosted, a defrost cycle is initiated. At this time the NC contacts open and the compressor and all fans are stopped. The NO contacts are closed, energizing the electric defrost heater and drain heater. In a short period of time, the contacts return to their normal operating position. The defrost cycle continues until the evaporator temperature reaches approximately 50 °F (10 °C). The defrost limiter control then terminates the defrost cycle and the unit operates in the normal cooling cycle.

*Defrost Limiter.* The defrost limiter used on domestic refrigerators is a temperature-sensing device that terminates the defrost cycle when the evaporator temperature reaches approximately 50 °F (10 °C). This control is fastened to the evaporator to sense the temperature, indicating that all the frost has been melted from the coil, and to terminate the defrost cycle. The defrost limiter has a set of NO contacts, which close on a fall in temperature.

*Operation.* When the timer contacts switch, indicating that a defrost period is needed, the compressor and all fans are stopped and the electric heating elements are energized. The defrost-limiter contacts are closed because of the lower temperature, and the unit goes into defrost. The electric heating element begins to warm up the evaporator and melt the frost from it. When the evaporator temperature reaches the cut-out temperature setting of the limiter, its contacts open and the system is returned to the cooling cycle.

# REVIEW QUESTIONS

1. On what types of systems are defrost systems used?
2. Define defrost systems.
3. What type of defrost system is the most economical?
4. What are the two main classifications of defrost systems?
5. What is the simplest type of defrost system?
6. What precaution should be taken with the condensate drain line on commercial refrigeration systems?
7. What control is used to terminate the defrost cycle?
8. What fan or fans are turned off during the defrost cycle?
9. At what time should the defrost cycle be initiated?
10. What objections are there to using a low-pressure control for a defrost control?
11. What is the most desirable length of defrost cycle?
12. What type of solenoid valve should be used in the hot-gas defrost system?
13. On what coil does the air-pressure-differential defrost system measure the pressure difference?
14. On the time-temperature type of defrost system, what is required to initiate a defrost cycle?
15. On what type of refrigeration systems are the compressor and fans stopped and electric resistance heaters energized during the defrost cycle?

# 17

# Service Hints
# On Controls

1. A control will cut out if
   a. The cut-out temperature setting is below the lower limit of the refrigerating system.
   b. Control contacts are short circuited by faulty wiring.
   c. Too much ice is on the chilling unit.
2. A control will not cut in if
   a. Power element has lost its charge.
   b. Open circuit exists in control wiring.

PILOT BURNER
SERVICE SUGGESTIONS

| TROUBLE | CAUSE | REMEDY |
|---|---|---|
| Pilot cannot be lighted | Pilot gas supply turned off | Turn on and relight pilot |
| | Pilot line contains air | Purge line |
| | Pilot burner orifice clogged | Clean and relight |
| | Lighting knob not being depressed | Depress knob and light |
| | Lighting knob not set at pilot position | Set knob and relight |
| | Pilot gas flow adjustment closed off | Readjust and light pilot |
| Pilot goes out when reset knob is released | Lighting knob released too soon | Hold knob in position longer |
| | Reset button released too soon | Hold reset in position longer |
| | Thermocouple or powerpile is bad | Replace |
| | Pilotstat power unit bad | Replace pilotstat |
| | Bad connection on power unit | Clean and tighten connection |
| | Pilot flame improper size | Adjust flame |
| | Powerpile terminals shorted or loose | Clean and tighten connections or replace |
| | Pilot filter clogged | Replace or clean filter |
| Pilot goes out when system is in use | Pilot burner orifice clogged | Remove and clean orifice |
| | Gas supply pressure too low | Check for restriction or adjust main gas pressure regulator |
| | Pilot unshielded from excessive draft | Shield pilot |
| | Thermocouple or powerpile bad | Replace |
| | Pilotstat power unit bad | Replace pilotstat |
| | Power unit connection dirty, loose, or wet | Clean, dry, and tighten properly |
| | Pilot flame improper size | Adjust flame |
| | Powerpile terminals shorted or loose | Repair terminals |
| | Thermocouple cold junction too hot | Reposition thermocouple |
| | Pilot burner lint screen clogged | Clean |
| | Pilot unshielded from burner concussion | Shield or adjust main burner to stop concussion |
| | Pilot filter clogged | Clean or replace filter |

## PILOT BURNER
## SERVICE SUGGESTIONS
*(cont.)*

| TROUBLE | CAUSE | REMEDY |
|---|---|---|
| | Gas supply too low | Increase gas pressure |
| | Pilot blowing away from thermocouple | Shield pilot from draft |
| | Bad thermocouple or powerpile | Replace thermocouple or powerpile |
| | Bad coil in pilotstat | Replace coil or pilotstat |
| Pilot burning but unit turned off by safety control | Power unit connection bad | Correct connection |
| | Pilot flame improper size | Adjust pilot flame or clean pilot burner |
| | Powerpile terminals shorted or loose | Correct condition of connections |
| | Improper venting | Correct vent problem |
| | Too small gas line, or line restricted | Increase line size or clear restriction |
| | Pilot burner lint screen clogged | Clean screen |
| | Pilot burner orifice to large | Replace orifice with proper size |
| Pilot flame wavering and yellow | Excessive ambient temperature | Change location of pilot burner or provide more air |
| | Excessive ambient temperature | Change location of pilot burner |
| | Improper venting | Correct venting problem |
| | Gas supply pressure too low | Increase gas pressure |
| | Pilot unshielded from combustion products | Relocate pilot burner or shield |
| Pilot flame small, blue, waving | Pilot burner orifice clogged | Remove and clean orifice |
| | Pilot gas flow adjustment is closed off | Open adjustment |
| | Pilot burner orifice too small | Replace orifice with correct size |
| | Pilot filter clogged | Replace or clean filter |
| Pilot flame noisy, lifting, blowing | Pilot gas pressure too high | Reduce gas pressure |
| | Pilot burner orifice too small | Replace orifice with correct size |
| Pilot flame hard, sharp | Typical of manufactured butane-air and propane-air mixture | No correction |
| Pilot turn-down test bad | Pilot burner improperly located | Relocate |

THERMOSTAT
SERVICE SUGGESTIONS

| TROUBLE | CAUSE | REMEDY |
|---|---|---|
| Thermostat jumpered; system still doesn't work | Thermostat not faulty | Check elsewhere |
| | Limit control set too low | Raise setting |
| | Low voltage control circuit open | Find open circuit and repair |
| | Low voltage transformer bad | Replace transformer |
| | Main gas valve bad | Replace valve |
| | Bad terminals | Repair terminals |
| System works when thermostat is jumpered | Dirty thermostat contacts | Clean contacts |
| | Damaged thermostat | Replace thermostat |
| Room temperature overshoots thermostat setting | Thermostat located on a cold wall | Change location |
| | Thermostat wiring hole not plugged | Plug hole with putty |
| | Thermostat exposed to cold draft | Relocate thermostat |
| | Thermostat not exposed to circulating air | Relocate thermostat |
| | Mercury type thermostat not mounted level | Level thermostat |
| | Thermostat not in calibration | Calibrate, if possible. If not, replace |
| | Anticipator set too high | Reset anticipator |
| | Heating plant too large | Reduce BTU input |
| | No anticipator in thermostat | Replace thermostat |
| Room temperature doesn't reach thermostat setting | Thermostat not mounted level | Level thermostat |
| | Thermostat not calibrated properly | Recalibrate or replace thermostat |
| | Heating plant too small or underfired | Increase BTU input |
| | Limit control set abnormally low | Set limit control |
| | Thermostat exposed to direct rays of the sun | Relocate thermostat |
| | Thermostat affected by fireplace or heat from appliances | Relocate thermostat |
| | Thermostat located on warm wall, or near a register | Relocate thermostat |
| | Dirty thermostat contacts | Clean contacts |
| | Bad wiring on terminals | Repair terminals |
| | Dirty air filter | Clean or replace filter |
| Thermostat cycles unit too often | Heat anticipator set too low | Reset to correct amperage |

## THERMOSTAT
## SERVICE SUGGESTIONS
*(cont.)*

| TROUBLE | CAUSE | REMEDY |
|---|---|---|
| Thermostat doesn't cycle unit often enough | Thermostat not exposed to return air | Relocate thermostat |
| | Too small heating plant or plant is underfired | Increase BTU input |
| | Heat anticipator set too high | Set to correct amperage |
| | No anticipator on thermostat | Replace thermostat |
| | Dirty thermostat contacts | Clean contacts |
| Too much variation in room temperature | Thermostat not exposed to return air | Relocate thermostat |
| | Heat anticipator set high | Set to correct amperage |
| | Heating unit too large or overfired | Reduce BTU input |
| | No heat anticipator | Replace thermostat |

## MOTOR SERVICE
## SUGGESTIONS

| TROUBLE | CAUSE | REMEDY |
|---|---|---|
| Fan motor will not run | Bad motor bearings, starting switch, or burnt winding | Repair or replace motor |
| | Fan control contacts not completing circuit | Replace fan control |
| | Fan relay contacts not completing circuit | Replace fan relay |
| | Blown fuse | Replace fuse |

## SOLENOID VALVE
## SERVICE SUGGESTIONS

| TROUBLE | CAUSE | REMEDY |
|---|---|---|
| Valve will not open | Thermostat or other controller inoperative | Repair or replace inoperative controls |
| | Clocks, limit controls, or other devices holding circuit open | Check circuit for limit control operation, blown fuses, short circuit, loose wiring, etc. |
| | Solenoid coil shorted or wrong voltage | Replace coil |
| Valve will not close | Manual opening device holding valve open | Release manual opening device |
| | Valve not mounted vertically | Mount valve in horizontal line with solenoid in vertical position above pipe |
| | Bent or nicked plunger tube restricting valve opening | Replace plunger tube or solenoid |
| | Foreign matter in valve interior | Disassemble valve and clean thoroughly |
| | Limit controls in grounded side of circuit | Rewire control into hot side of circuit |

### DEFROST SYSTEMS
### SERVICE SUGGESTIONS
(Electric heat or hot gas)

| TROUBLE | CAUSE | REMEDY |
|---|---|---|
| Defrosts at wrong time | Defrost control set on wrong time | Set knob to indicate proper time of day |
| | Control may defrost at wrong knob position | Replace the control |
| | Timer motor current different than power supply | Replace control to correspond to power supply |
| Does not keep time | Clock does not run | Check for voltage at motor terminals; if power is found replace control. If not, check elsewhere |
| | Intermittent power supply or wrong wiring connections | Check power supply or correct connections |
| Will not defrost | Clock does not run | Check for voltage at motor terminals; if voltage is found replace control. If not, check elsewhere |
| | Clock runs but does not turn dial shaft | Replace control |
| | Defrost control bulb too warm | Properly locate and clamp control bulb |
| | Skipped a defrost cycle | Replace control if trouble persists |
| | Leak in fail-safe bellows | Replace control |
| Incomplete defrost | Excessive frost build-up between defrost periods | Check defrost operation by turning knob to just before defrost. Determine and correct excessive build-up. (Door may be loose or left open too much, uncovered foods, etc.) |
| | Defrost termination switch set too low | Raise temperature setting, (each 1,000 ft altitude lowers setting about 2.5° F) |
| | Partial lost fill of fail-safe bellows | If recalibration does not correct, replace control |
| | Bulb not properly attached to evaporator | Clamp bulb in clean area |
| | Cross ambient condition | Control must be warmer than bulb or tubing during operation |
| | Compressor cycles on overload | Replace starting relay or overload |
| | Defective defrost switch | Check defrost control contacts; if open replace control |
| | Cycling control lost fill | Check cycling control; if contacts are open replace control |

| TROUBLE | CAUSE | REMEDY |
|---|---|---|
| Defrost will not terminate | Cross ambient condition | Be sure control is warmer than the bulb or tubing |
| | Solenoid valve stuck open | If no voltage is on coil, replace solenoid valve |
| | Lost fill of non-fail-safe bellows | Replace control |
| | Termination setting too high | Correct setting |
| Noisy control | Control mounted in wrong position | Correct mounting position; if still noisy replace control |
| | Noisy clock motors | Replace control |

## OIL BURNER
## SERVICE SUGGESTIONS

| TROUBLE | CAUSE | REMEDY |
|---|---|---|
| Repeated safety shutdown | Inadequate combusion detector response | If no response, replace detector. If response is inadequate, move detector to respond to stable part of flame |
| | Low line voltage | New wiring, or contact power company |
| Short cycling of burner | Dirty filters | Clean |
| | Faulty operation of auxiliary controls | Reset, repair, or replace auxiliary controls |
| | Incorrect thermostat anticipation | Set to higher amperage |
| Relay will not pull in | No power. Open power circuit | Reset, repair, or replace |
| | Open thermostat circuit | Check thermostat wiring |

# Electric Controls for Refrigeration and Air Conditioning

## Workbook

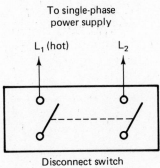

Disconnect switch

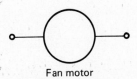

Fan motor

Connect the power source through the disconnect switch to the fan motor so that it will operate when the switch is closed.

To three-phase
power supply

$L_1$　　$L_2$　　$L_3$

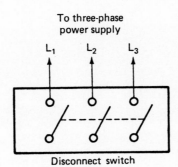

Disconnect switch

Compressor motor

Connect the power source through the disconnect switch to the compressor motor so that it will operate when the switch is closed.

To single-phase
power supply

L$_1$ (hot)              L$_2$

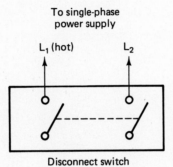

Disconnect switch

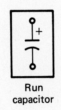

Run
capacitor

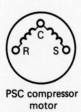

PSC compressor
motor

Connect the power source through the disconnect switch to the permanent split-capacitor compressor motor so that it will operate when the switch is closed.

To power supply

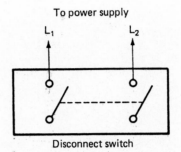

Disconnect switch

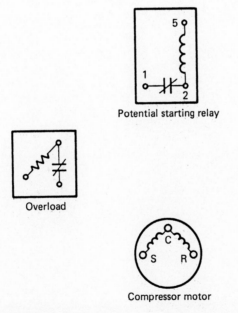

Potential starting relay

Overload

Compressor motor

Complete the diagram to include the starting relay and overload.

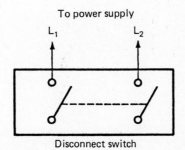

To power supply

Disconnect switch

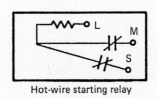

Hot-wire starting relay

Compressor motor

Complete the wiring diagram including the hot wire relay.

To power supply

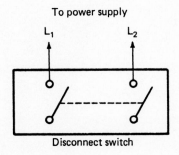

Disconnect switch

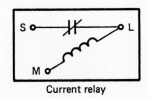

Current relay

Overload

Compressor motor

**Complete the diagram including all components.**

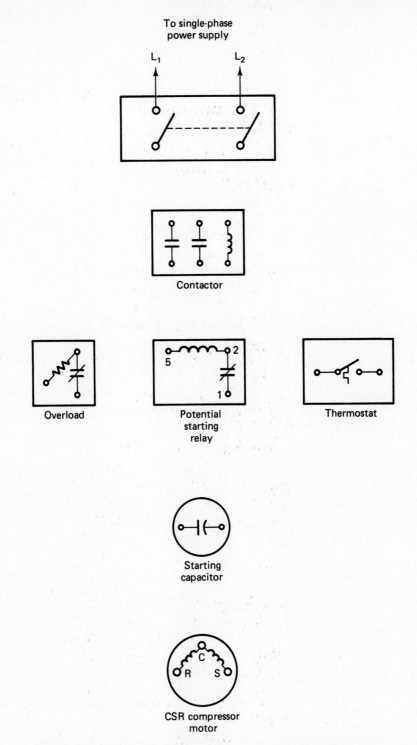

To single-phase
power supply

L₁          L₂

Contactor

Overload

Potential
starting
relay

Thermostat

Starting
capacitor

CSR compressor
motor

Complete the diagram including all components in the line voltage
circuit.

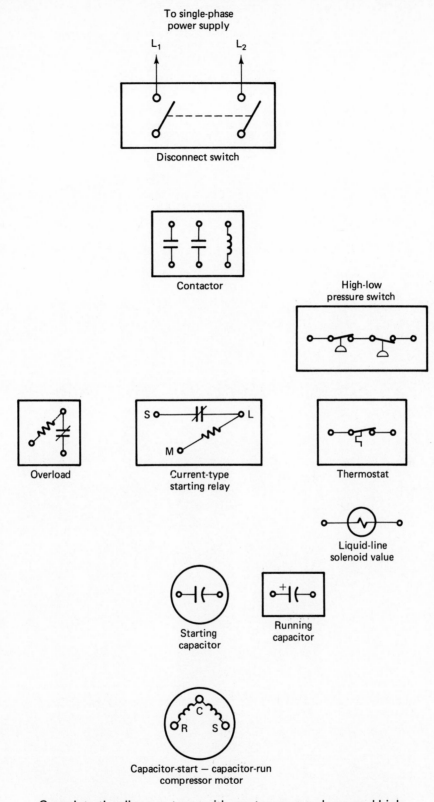

To single-phase
power supply

L₁          L₂

Disconnect switch

Contactor

High-low
pressure switch

Overload

Current-type
starting relay

Thermostat

Liquid-line
solenoid value

Starting
capacitor

Running
capacitor

Capacitor-start — capacitor-run
compressor motor

Complete the diagram to provide system pump-down and high-pressure safety.

To three-phase
power supply

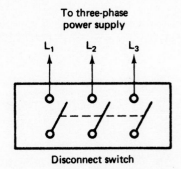

Disconnect switch

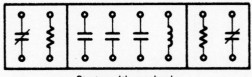

Starter with overloads

High-low
pressure switch

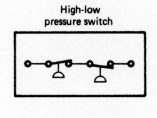

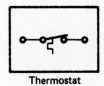

Thermostat

Liquid-line
solenoid value

Three-phase
compressor motor

Complete the diagram to provide system pump-down and high-pressure safety. Include all components.

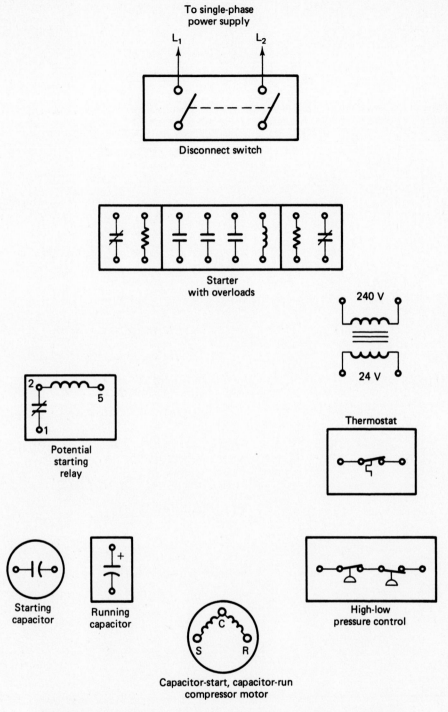

To single-phase
power supply

L₁          L₂

Disconnect switch

Starter
with overloads

240 V

24 V

Potential
starting
relay

Thermostat

Starting
capacitor

Running
capacitor

Capacitor-start, capacitor-run
compressor motor

High-low
pressure control

**Complete the diagram, placing all controls possible in the low-voltage circuit.**

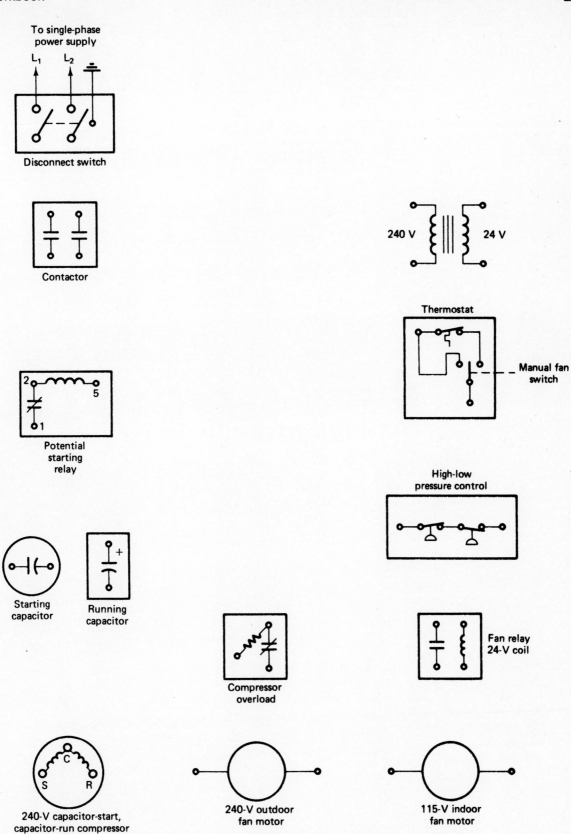

To single-phase power supply

L₁   L₂

Disconnect switch

Contactor

240 V   24 V

2   5
1

Potential starting relay

Thermostat

Manual fan switch

High-low pressure control

Starting capacitor

Running capacitor

Compressor overload

Fan relay 24-V coil

C
S   R

240-V capacitor-start, capacitor-run compressor motor

240-V outdoor fan motor

115-V indoor fan motor

Complete the cooling diagram, placing all control devices in the low-voltage (24-V) circuit. Use a red pencil for voltage, a green pencil for the fan circuit, and a yellow pencil for cooling.

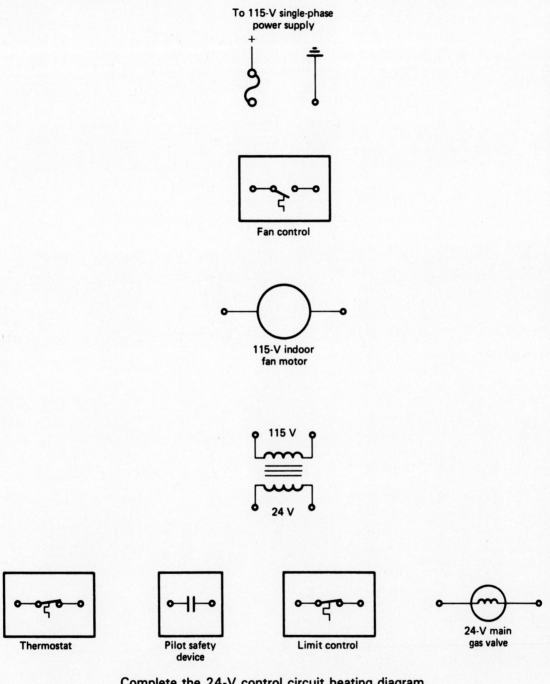

To 115-V single-phase
power supply

Fan control

115-V indoor
fan motor

115 V

24 V

Thermostat        Pilot safety       Limit control        24-V main
                  device                                   gas valve

Complete the 24-V control circuit heating diagram.

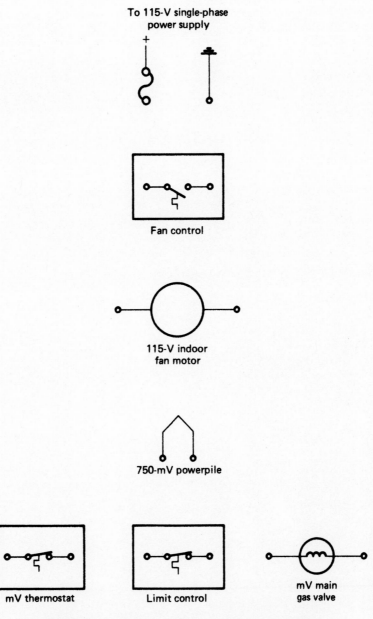

To 115-V single-phase
power supply

Fan control

115-V indoor
fan motor

750-mV powerpile

mV thermostat          Limit control              mV main
                                                  gas valve

Complete the above 115V-mV heating diagram.

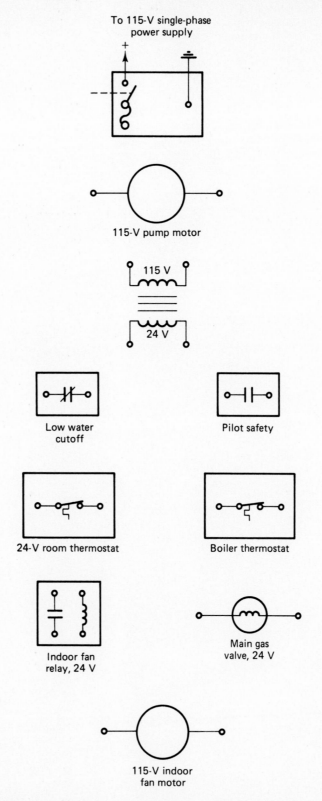

To 115-V single-phase
power supply

115-V pump motor

115 V

24 V

Low water
cutoff

Pilot safety

24-V room thermostat

Boiler thermostat

Indoor fan
relay, 24 V

Main gas
valve, 24 V

115-V indoor
fan motor

Complete the diagram of a hot-water boiler so that the indoor
24-V thermostat controls the indoor fan and the boiler thermostat controls the gas valve through the safety devices.

To 115-V single-phase
power supply

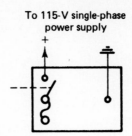

Low-water
cutoff

Pilot safety
control

High-water
cutoff

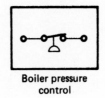

Boiler pressure
control

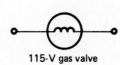

115-V gas valve

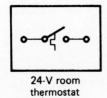

24-V room
thermostat

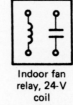

Indoor fan
relay, 24-V
coil

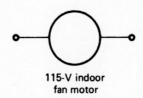

115-V indoor
fan motor

Complete the diagram so that the boiler controls and indoor controls do not affect each other.

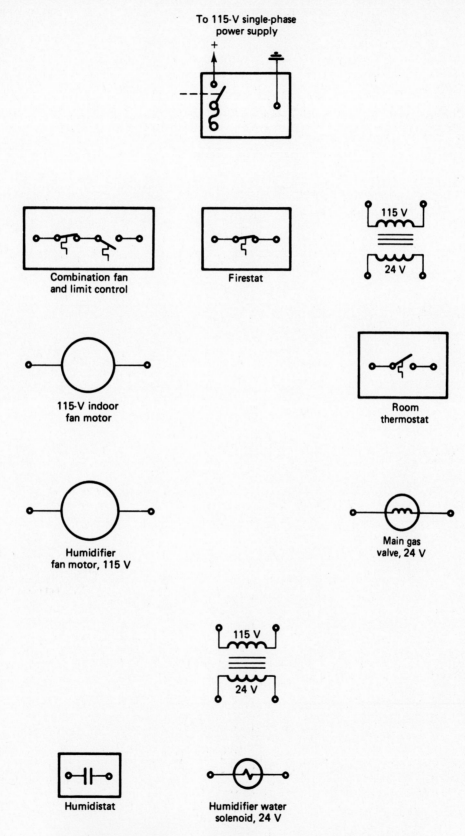

To 115-V single-phase power supply

Combination fan and limit control

Firestat

115 V
24 V

115-V indoor fan motor

Room thermostat

Humidifier fan motor, 115 V

Main gas valve, 24 V

115 V
24 V

Humidistat

Humidifier water solenoid, 24 V

Complete the diagram so that the humidifier will operate only when the indoor fan is operating.

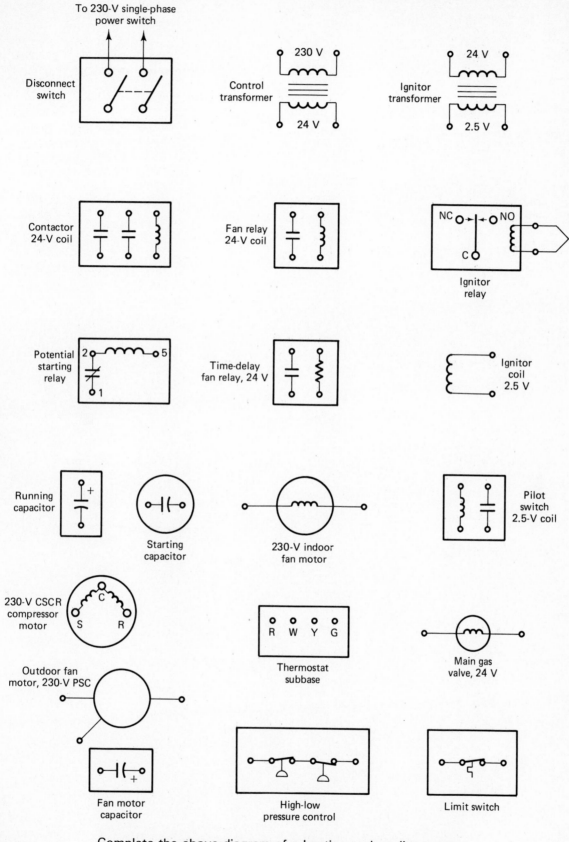

Complete the above diagram of a heating and cooling system with an automatic pilot ignitor. Use colored map pencils: Red-voltage; White-heating; Yellow-cooling; Green-fan; Brown-fan capacitor.

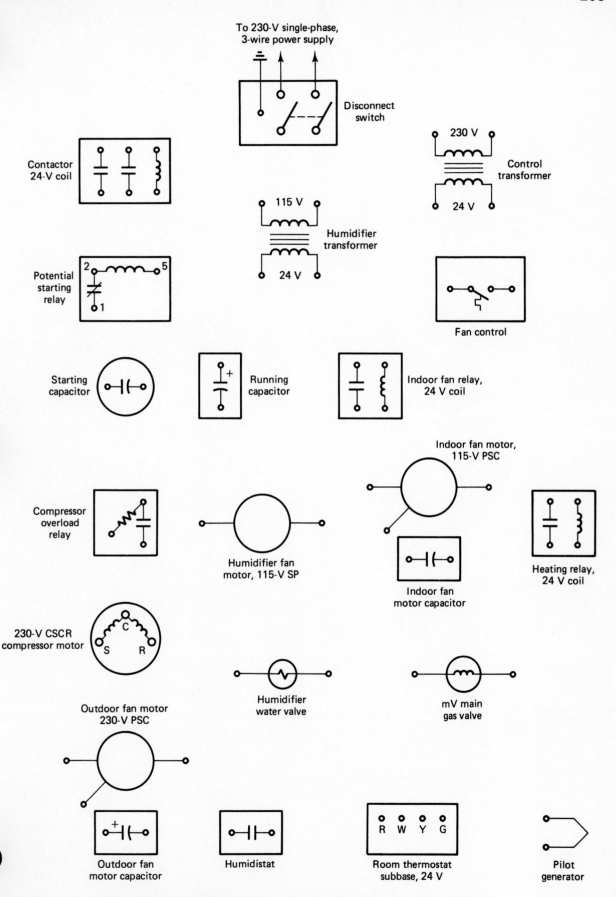

Complete the above wiring diagram, following the color coding
as set before. Do not mix the voltages.

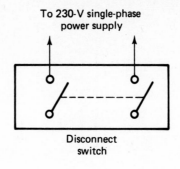

Disconnect
switch

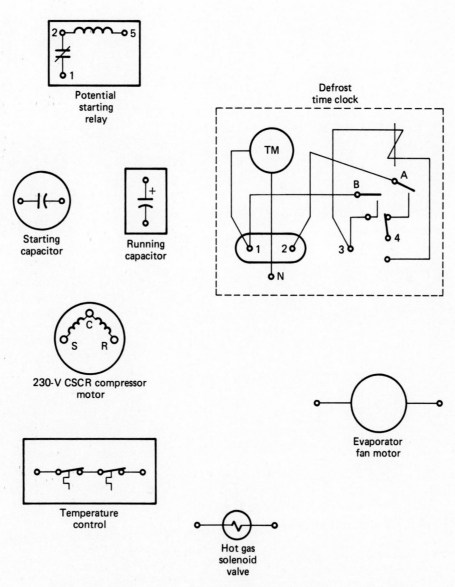

Complete the diagram of a simple hot-gas defrost system.

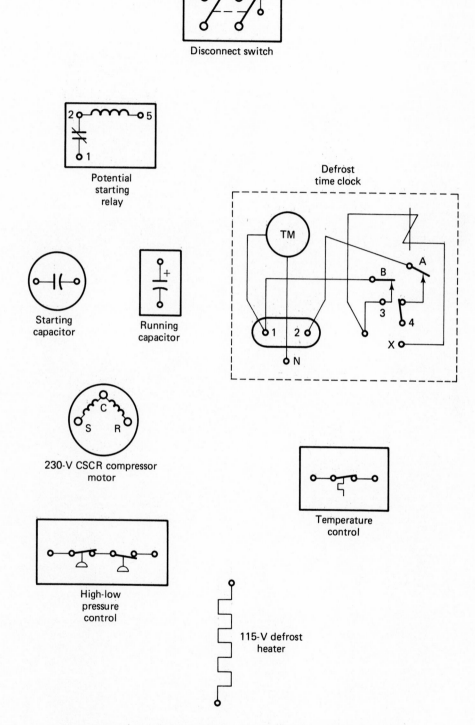

To 230-V single-phase
power supply

Disconnect switch

Potential
starting
relay

Starting
capacitor

Running
capacitor

Defrost
time clock

TM

230-V CSCR compressor
motor

Temperature
control

High-low
pressure
control

115-V defrost
heater

Complete the above diagram of an electric defrost system.

To 230-V single-phase
power supply

Disconnect switch

Contactor
no. 1, 24-V coil

115 V

24 V

Contactor
no. 2, 24-V coil

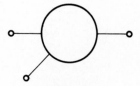

Indoor fan
relay, 24-V coil

Heating relay-
no. 1, 24-V coil

115-V PSC indoor
fan motor

Heating relay-
no. 2, 24-V coil

Indoor fan
motor capacitor

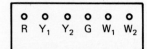

R   Y₁   Y₂   G   W₁   W₂

Two-stage-heat–two-stage-cool
thermostat subbase, 24 V

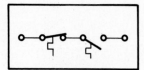

Combination
fan and limit
control

Complete the diagram for two-stage heating and two-stage cooling. Use the color code.

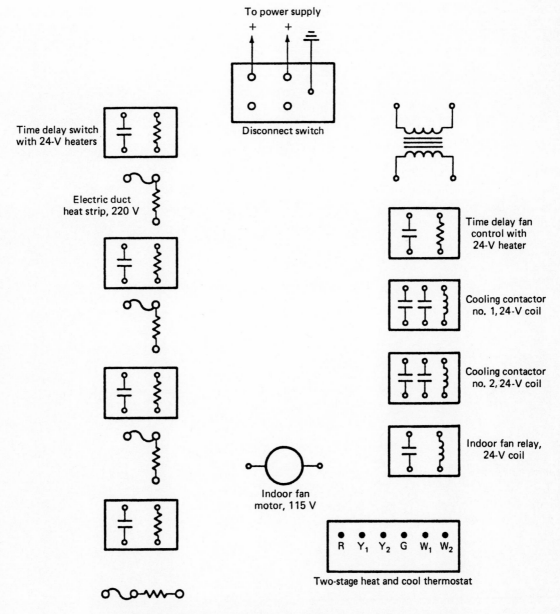

Time delay switch
with 24-V heaters

Electric duct
heat strip, 220 V

To power supply

Disconnect switch

Time delay fan
control with
24-V heater

Cooling contactor
no. 1, 24-V coil

Cooling contactor
no. 2, 24-V coil

Indoor fan relay,
24-V coil

Indoor fan
motor, 115 V

R   Y₁   Y₂   G   W₁   W₂

Two-stage heat and cool thermostat

Complete the diagram, wiring one-half the heat strips on each
thermostat stage. Use the color code.

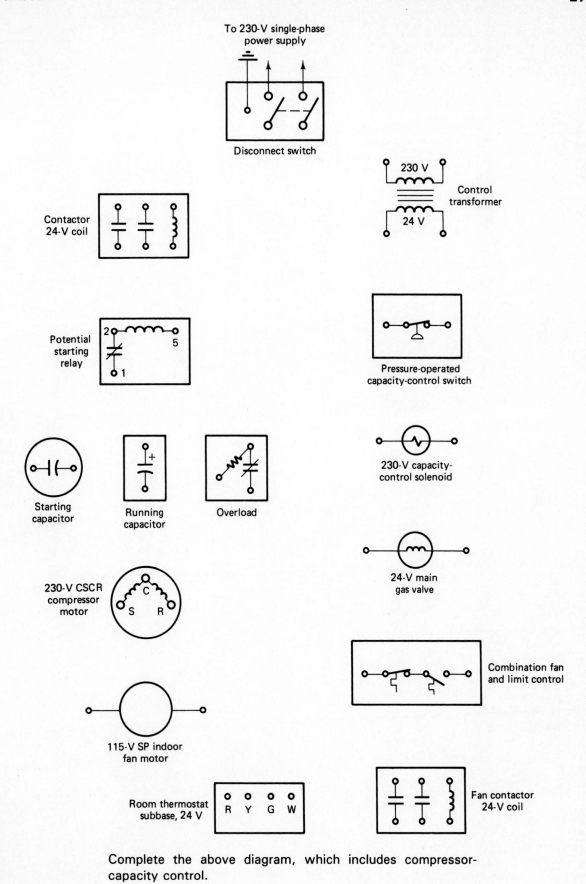

To 230-V single-phase power supply

Disconnect switch

Contactor 24-V coil

Control transformer
230 V
24 V

Potential starting relay

Pressure-operated capacity-control switch

Starting capacitor

Running capacitor

Overload

230-V capacity-control solenoid

24-V main gas valve

230-V CSCR compressor motor

Combination fan and limit control

115-V SP indoor fan motor

Room thermostat subbase, 24 V
R  Y  G  W

Fan contactor 24-V coil

Complete the above diagram, which includes compressor-capacity control.

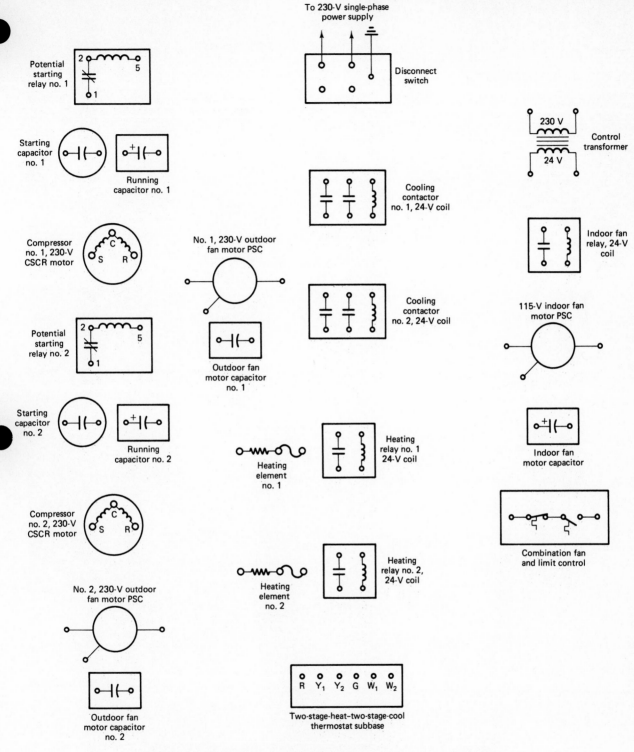

Complete the diagram for two-stage heating and two-stage cooling. Use the color code.

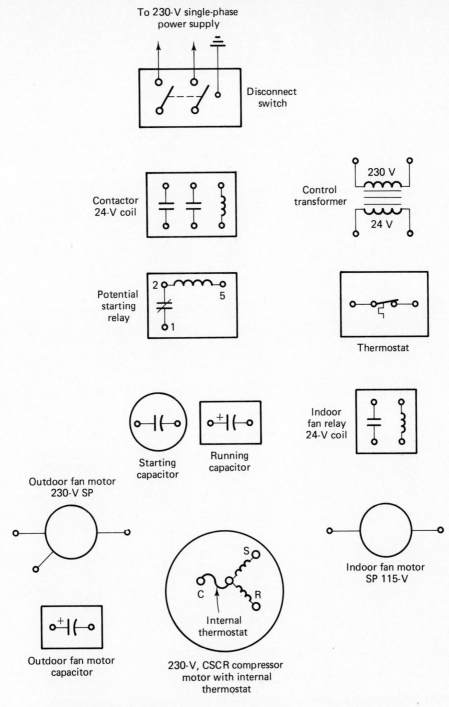

To 230-V single-phase power supply

Disconnect switch

Contactor 24-V coil

Control transformer

230 V

24 V

Potential starting relay

Thermostat

Starting capacitor

Running capacitor

Indoor fan relay 24-V coil

Outdoor fan motor 230-V SP

Indoor fan motor SP 115-V

Outdoor fan motor capacitor

230-V, CSCR compressor motor with internal thermostat

Internal thermostat

Complete the wiring diagram for the cooling system. Compressor has internal overload.

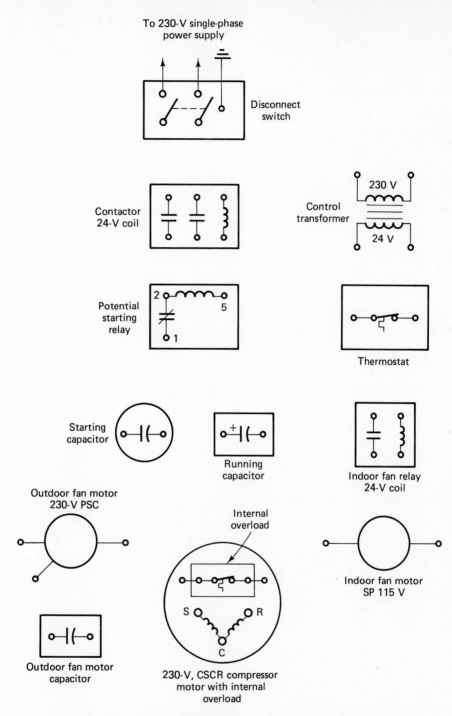

To 230-V single-phase
power supply

Disconnect
switch

Contactor
24-V coil

Control
transformer

230 V

24 V

Potential
starting
relay

2

5

1

Thermostat

Starting
capacitor

Running
capacitor

Indoor fan relay
24-V coil

Outdoor fan motor
230-V PSC

Internal
overload

S        R

C

Indoor fan motor
SP 115 V

Outdoor fan motor
capacitor

230-V, CSCR compressor
motor with internal
overload

Complete the wiring diagram for the indicated cooling system.
The internal thermostat controls the contactor.

# Glossary

**Adjustable Differential:**  A means of changing the difference between the control cut-in and cut-out settings.

**Ambient Compensated:**  Such a control is designed so that varying air temperature at the control does not materially affect the control setting.

**Ambient Temperature:**  Temperature of the surrounding air in the immediate vicinity of the control body and power element system.

**Ampacity:**  The current capacity of the electrical conductor, depending upon the type of material it is made from, the conductor size, and the type of insulation on the conductor.

**Amps:**  The measure of electrical current flowing through an electrical circuit.

**Aquastat:**  A control used on hydronic heating-cooling systems to measure the temperature of the fluid in a pipe or boiler and make a change in the operation of the controlled equipment.

**Automatic Changeover:**  The automatic change from one function to another. Otherwise, changes would be accomplished manually.

**Automatic Flue Damper:**  A damper that is placed at the outlet of the draft diverter on gas-burning equipment, which opens when operation of the equipment is required and closes when the requirement has been met. Dampers are used to conserve gas by preventing the venting of flue gases from the equipment during the off cycle.

**Automatic Recycle:**  Control contacts return to the original position automatically after actuation when the pressure or temperature returns to normal condition.

**Auxiliary Contacts:**   A set of contacts used to perform a secondary function, usually in relation to the operation of the main contacts—for example, to cause an alarm to operate if the main contacts actuate.

**Bellows:**   A corrugated metallic diaphragm with a metal cup. Often a complete power unit is referred to as a bellows.

**Bimetal:**   Two pieces of metal welded together so that they function as one piece. They are used in thermostats, thermometers, and other heat-sensing devices. A bimetal will bend in one direction when heated and return to the original position when cooled to the original temperature.

**Boiler Pressure Control:**   A control used to operate a boiler burner in response to the pressure inside the system. It is used to cycle the burner on and off as the pressure rises and falls.

**Blower Control:**   A relay used for the automatic operation of an indoor blower in response to the requirements of another circuit. The contacts are closed on demand from the temperature-control circuit.

**Break Point:**   Temperature at which all the refrigerant charge of the element has completely vaporized.

At temperatures above the break point of the fill, there is a very little internal pressure change in the element—approximately 1 pound per square inch gauge (6.89 kilopascals) for each 10°F (5.56°C) temperature change.

**British Thermal Unit:**   The amount of heat required to raise the temperature of 1 pound (0.4536 kilogram) of water 1°F (0.56°C).

**Capacitor:**   A device used to boost the voltage to a motor. Running capacitors are used in the starting winding to increase the running torque of the motor. Starting capacitors are used in the starting winding to increase the starting torque of the motor.

**Coil:**   A coil of wire placed in close relation to a movable armature, which when energized electrically causes the armature to move in the desired direction. They are used in every relay, contactor, and starter in electrical systems.

**Combination Fan and Limit Control:**   A control that cycles the fan and the main burner off and on in response to the temperature of a designated place in the heating equipment.

**Compressor Unloaders:**   Devices placed on the compressor to reduce or increase its pumping capacity in response to the system requirements. They generally operate in response to the suction pressure of the system.

**Contactor:**   A device used to energize a heavy electrical load such as a compressor or heating element in an electric heating unit. They are usually actuated from a different circuit than that being controlled.

**Contact Rating:**   The capacity of electrical contacts to handle current and voltage.

**Cooling Anticipator:**   A resistor placed inside the thermostat in electrical parallel to the cooling temperature control circuit. It causes the cooling unit to cycle on before the temperature actually rises to the cut-in setting of the thermostat.

**Cross-ambient Fill:**   A vapor-pressure element sufficiently large to assure liquid in the bulb regardless of whether the bulb is colder or warmer than the control ambient temperature. This occurs when temperatures at the control and capillary may be alternately above and below the desired temperature at the sensing bulb.

**Current Relay:**   A switching relay that operates on a predetermined amount of electrical flow (or lack of current flow).

**Cut-in Setting:**   The point at which the control electrical contacts close to make an electrical circuit.

**Cut-out Setting:**   The point at which the control electrical contacts open to break an electrical circuit.

**Defrost Control:**   The control used to determine when a system needs defrosting. It initiates and in some cases terminates the defrost cycle on heat pump and commercial refrigeration systems.

**Differential:**   The difference between the cut-in and cut-out settings of a control.

**Differential Screw:**   An adjusting screw used to change the difference between the cut-in and cut-out settings of a control.

**Direct Spark Ignition:**   A method by which the main-burner gas is ignited by a spark across the ports of the main burner rather than lighting a pilot burner. The spark is generated by a high-voltage transformer in the ignition circuit.

**Double Pole:**   Two single-pole contacts operating simultaneously.

**Double Throw:**   Contacts make in one direction but break simultaneously in the other direction.

**Drop-out Voltage:**   The voltage or point at which the pull of the electromagnet is not strong enough to keep the armature seated.

**Dry-bulb Temperature:**   The temperature indicated on a dry bulb thermometer. It indicates the heat of the air and water vapor mixture.

**Dummy Terminals:**   Extra terminals that do not connect to an electrical function on a control. They are sometimes provided for wiring convenience.

**Enthalpy:**   The total heat contained within a quantity of a substance. This includes both the sensible and the latent heat of the substance.

**Fail-safe Control:**   A control designed such that a component failure will cause the control to assume the safest action, thus protecting the system on which it is installed, usually a contact-open condition.

**Fan Control:**   A switch used to cycle the indoor fan on and off in response to the temperature inside the heat exchanger. The control is usually set to start the fan at approximately 135 °F (57 °C) and stop it at approximately 100 °F (37.8 °C).

**Fixed Differential:**   The factory-set difference between the control cut-in and cut-out setting. It cannot be changed.

**Fixed Setting:**   Provides no convenient means for changing control of fixed settings after the control leaves the factory.

**Flame Rectification:** A method for operating flame safety equipment by ionizing an area by a pilot flame. This flame completes an electrical circuit in the safety circuit and allows the equipment to operate.

**Flux:** The electric or magnetic lines of force in a region.

**Full-load Amps:** The amount of current in amps in an electrical circuit when the load is operating in a full-capacity condition.

**Glow Coil:** A coil caused to become red hot by electrical current flowing through it to light the pilot burner gas on automatic-pilot ignition systems.

**Ground:** Intentional or accidental connection from a power source to the earth, or a connecting body that serves in place of the earth to complete an electrical circuit. Earth is considered zero potential.

**Hard-start Kit:** A group of components used to increase the starting torque of an electric motor. It consists of a starting capacitor, a relay, and the necessary wiring to install the kit on the unit.

**Heating Anticipator:** A resistor located inside the thermostat and placed in series with the temperature-control circuit during the heating cycle to cause a false heat inside the thermostat. This causes the thermostat to stop the heating equipment before the temperature inside the space actually reaches the set point.

**Heat Pump:** An electrically operated device designed to extract heat from one location and transfer this heat to another location. A heat pump is used for both heating and cooling and reverses the refrigeration cycle when heating is needed.

**Hermetically Sealed:** A compressor or other device enclosed in a gas-tight housing.

**Hertz (hz):** The frequency in cycles per second of an ac power source. In the United States this power is generally 60 hertz.

**High-pressure Control:** A control designed to sense the discharge pressure of a compressor and to stop the compressor when this pressure reaches an unsafe point to prevent damage to the compressor motor.

**Horsepower Rating:** A rating in terms of a motor. Underwriters' Laboratories consider 1 horsepower equivalent to 746 watts. Most ratings now refer to electrical ratings in amps.

**Hot-wire Relay:** A relay that is designed to help start a split-phase compressor motor by directing electric current to the start winding. The relay removes the start winding from the circuit by causing a bimetal switch to warp open because of the current flowing through the relay.

**Humidity Controller:** A control designed to start and stop humidifying equipment in response to the amount of relative humidity inside the controlled space.

**Ignitor-sensor:** Flame-safe equipment designed to ignite the pilot-burner gas and then sense if there is a flame present. If no flame is present or if the flame is not satisfactory to ignite the main burner gas, the control system will not open the main gas valve.

**Inductive Rating:** The maximum amount of amperes in a circuit when a conductor is in an electromagnetic field.

**Inherent Motor Protection:** A safety-limit device built inside a motor or equipment. It protects for overtemperature, overcurrent, or both.

**Intermittent Pilot:** A pilot lit each time the thermostat demands heat. When the thermostat is satisfied, the pilot light and main burner are turned off at the same time.

**Internal Overload:** A protective device placed inside the motor winding at a predetermined place to protect the motor from overheat, overcurrent, or both.

**Limit Control:** A control mounted inside the heat exchanger to sense an overtemperature condition. If an overtemperature condition should be sensed, the main-burner gas will be extinguished to prevent a possible fire or damage to the equipment.

**Locked-rotor Amps (LRA):** Current required at the instant power is supplied to start the motor.

**Low-ambient Kit:** A kit consisting of the required components to cycle the condenser fan to maintain the head pressure high enough so that the equipment will operate satisfactorily.

**Low-pressure Control:** A control designed to stop the compressor if the suction pressure drops to a predetermined point. The control contacts are generally in the control circuit, but on some cases they may be in the line voltage to the compressor motor.

**Low Water Cutoff:** A control designed to stop the main burner on a boiler should the water inside the system drop to a dangerous level.

**Manual Reset:** A control-reset mechanism that requires a manual operation if the control locks out on safety.

**Modulating Control:** A control in which corrective action is in small increments, as opposed to complete on-off action.

**Normally Closed (NC):** A switch or a valve that remains closed when the device is not connected to a power supply or is de-energized.

**Ohm's Law:** The basic relationship between the voltage ($E$), the current ($I$), and the resistance ($R$) of an electric circuit.

**Oil-failure Control:** A control that is designed to stop the compressor should the oil pressure drop to a dangerously low point to prevent mechanical damage to the compressor.

**On-off:** Used to control operations. The system is either on or off (two positions) as opposed to proportional or modulating.

**Open Circuit:** An electrical circuit without continuous path for current to flow. This open circuit may be caused by an open switch or a broken circuit such as a blown fuse.

**Outdoor Reset Control:** A control designed to sense both the indoor and outdoor air temperature and change the operating cycle of the equipment in response to the demands of the space being treated.

**Overload Protector:** A device that opens ungrounded conductors for protection against motor overcurrent. It prevents unsafe running conditions and protects the motor from burnout.

**Pilot Burner:** A burner designed to light the main burner on heating equipment. It is sometimes used to provide the heat required to operate the pilot safety devices used on gas burning equipment.

**Pilot-duty Rating:** An electrical rating applied to devices used to energize and de-energize pilot circuits, such as the holding coil of a motor contactor.

**Pilot Safety:** A device used to prevent unsafe operation of the main burner should the pilot burner flame not be sufficient to ignite the main burner gas safely.

**Positive-temperature-coefficient Starting Device:** A compressor-starting device that increases in resistance with an increase in temperature. When the resistance is increased to a predetermined level, the current flow is reduced to a small trickle, effectively removing the starting components from the system.

**Potential Relay:** A relay used to start motors requiring a high starting torque. The relay coil is energized by a voltage generated in the starting winding of the motor.

**Power:** The voltage used for actuating a device. A common use of power refers to electrical power or voltage.

**Proportional Controls:** See modulating.

**Protectorelay:** A device used on oil burners to provide safety ignition of the oil on start-up. It is a heat-sensing control mounted on the unit stack to sense heat coming from the flame. If no heat is detected, the protectorelay will shut the unit down.

**Pull-in Voltage:** The voltage value that causes the relay armature to seat on the pole face.

**Quick Connects:** Terminals of a switch that are usually connected by a pushing motion rather than the normal screw terminals.

**Range:** The pressure- or temperature-operating limits of a control.

**Range-adjusting Screw:** An adjusting screw used to change the operating set points of a control. Changes are limited to those within the control range.

**Redundant Gas Valve:** A valve that has two seats, both of which must be open before the gas can flow through. However, it takes only one seat closed to stop the flow of gas. It is a safety device to prevent equipment overheating due to a gas valve stuck open.

**Relative Humidity:** Water vapor contained in a body of air as a percentage of the maximum water vapor density possible, which is 100% relative humidity.

**Relay:** An electromagnetically operated control that is used for the switching of electrical circuits with one circuit controlling another circuit, usually of higher voltage and current usage.

**Reversing (Four-way) Valve:** A valve used on heat-pump systems and commercial refrigeration defrost systems to reverse the flow of refrigerant through the system directing the hot gas to the evaporator to remove any ice accumulation.

**Run Winding:** The winding in an electric motor that provides the power for turning the rotor during operation. It has the larger wire of the two windings.

**Sequencer:** A device designed to start or stop pieces of equipment in a predetermined sequence. It is usually used on multistage systems for capacity control.

**Set Point:** Setting at which the desired control action occurs.

**Single Pole:** One set of two electrical contacts. These two contacts make and break on switch action.

**Single Throw:** Contacts make or break in only one direction of operation.

**Solenoid Valve:** A valve used to open and close a line. They are generally used in pump-down systems.

**Solid-state Start Relay:** A compressor-starting device that makes use of solid-state components to start a compressor motor.

**Standing Pilot:** A pilot burner that remains lighted after it has been lighted regardless of the system demands. It does not cycle with system demands.

**Starter:** A device used for the switching of heavy current, voltage, or both from another circuit. They are commonly used to start compressor motors and other heavy current using devices.

**Starting Relay:** A relay used to direct the electric current to the auxiliary winding in an electric motor during the starting period.

**Start Windings:** The auxiliary windings in an electric motor used to provide the extra torque for starting and to aid in providing the torque when the motor is running.

**Step Controller:** See Sequencer.

**Terminal:** An electrical connection, such as a screw terminal.

**Terminate:** To complete an event or stop an operation.

**Thermal Relay:** A relay actuated by the heating effect of an electrical current, which is sometimes referred to as a *warp switch*.

**Thermocouple:** A device made from two different metals and used to provide the electrical power to operate the pilot safety circuit in a gas-heating system.

**Thermopilot Relay:** A relay whose coil is energized by the electrical power generated by the thermocouple. Its contacts are in the temperature-control circuit from the thermostat.

**Thermostat:** A temperature-sensing device used to control the operation of heating and cooling equipment in response to the temperature of the conditioned space.

**Time Clock:** A timing device used to control the off- and on-cycles of equipment in response to the time of day. They have contacts that make and break circuits at a given time each day without personal attention.

**Time-delay Device:** Designed to provide a time interval between operations of a device.

**Underwriters' Laboratories:** A testing agency whose primary function is to assure that products are manufactured to meet specific safety standards. A listing of a product by the Underwriters' Laboratories indicates that the product was tested and met the recognized safety requirements.

**Voltage Relay:** See *potential relay*.

**Warp Switch:** A thermal relay switch actuated by the heating effect of an electric current.

**Water-regulating Valve:** A valve designed to control automatically the amount of water flowing through a water-cooled condenser to maintain a predetermined head pressure.

**Wet-bulb Temperature:** The temperature indicated on a wet-bulb thermometer. It indicates the total amount of heat in a mixture of air and water vapor.

# Index